MW00843359

WHITE LIGHT

WHITE LIGHT

The Elemental Role of Phosphorus—in Our Cells, in Our Food, and in Our World

JACK LOHMANN

PANTHEON · NEW YORK

FIRST HARDCOVER EDITION PUBLISHED BY PANTHEON BOOKS 2025

Published by Pantheon Books, a division of Penguin Random House LLC,
1745 Broadway, New York, NY 10019.

Pantheon Books and colophon are registered trademarks of
Penguin Random House LLC.

Illustrations by Alice Maiden

Library of Congress Cataloging-in-Publication Data
Names: Lohmann, Jack, author.
Title: White light : the afterlives of phosphorus / Jack Lohmann.
Description: New York : Pantheon, 2025. | Includes bibliographical references and index.
Identifiers: LCCN 2024009567 | ISBN 9780593316610 (hardcover) | ISBN 9780593316627 (ebook)
Subjects: LCSH: Phosphorus in agriculture—History—Popular works. | Phosphatic fertilizers—History—Popular works. | Phosphates—Environmental aspects—Popular works. | Phosphorus—Environmental aspects—Popular works. | Phosphates—Popular works. | Phosphorus—Popular works.
Classification: LCC S587.5.P56 L64 2025 |
DDC 631.8/5—dc23/eng/20241029
LC record available at https://lccn.loc.gov/2024009567

penguinrandomhouse.com | pantheonbooks.com

Printed in the United States of America
1 3 5 7 9 10 8 6 4 2

The authorized representative in the EU for product safety and compliance is Penguin Random House Ireland, Morrison Chambers, 32 Nassau Street, Dublin D02 YH68, Ireland, https://eu-contact.penguin.ie.

To my parents

CONTENTS

PROLOGUE

WHALE FALL

In the moments that follow the death of a whale, when the light disappears and is swallowed by dark, the body's weight draws to the base of the sea and compresses. It settles in mud. It forms an environment known as a whale fall, a world that will last for decades.

The whale fall grows in stages. The larger species come, the eels, the sharks. They rip apart the dead whale's flesh. The tail, the head, the organs are consumed. The size of predator lessens as the length of time extends. Tiny mouths clean the bones dry. A skeleton remains; bacteria descend upon it. They turn bones into nutrition, consuming the whale in a process that is almost imperceptibly slow. Worms arrive and burrow through the skeleton. Other organisms come and eat the worms. Larger predators reinhabit the space. Within a barren, lightless plain, on the basis of decaying bones, a world is born.

Whalebone contains an element that is rare: phosphorus, a limiting ingredient in life on earth. Of all the elements of the periodic table, phosphorus is one of six that are absolutely necessary for the existence of life. Of those six, phosphorus is the most limited. Because of its rarity, it controls life—it determines who grows and shrinks, who lives and dies, what areas become biologically wealthy and which ones will be biologically poor. "The maximum mass of protoplasm which the land can support, like the maximum that the sea can support, is dictated by the phosphorus content," Isaac Asimov, the biochemist, wrote in 1959. Phosphorus, he wrote, "is life's bottleneck."

Each of the six essential elements performs a vital role. Carbon forms long chains, connecting compounds together to create large, complicated structures. Hydrogen and oxygen combine to form water. Nitrogen and sulfur create proteins, providing organisms with food. Phosphorus converts energy, carries information, constructs cell membranes, and performs a host of other actions that underpin life's complexity. Phosphorus allows seeds to grow and fruit to ripen. It is the main ingredient in matches. It both enables life and destroys it. Sarin gas, created from white phosphorus, is a potent agent of chemical warfare favored in Syria today and in Saddam Hussein's Iraq. A drop of sarin can lead to death. White phosphorus itself is a chemical weapon, a notorious substance sometimes used in contravention of the law—notably by the governments of Syria, Israel, and the United States, and by the Taliban. On those rare occasions when white phosphorus has been deployed, it has fallen through the air in a deluge of dissipated fire.

When it is isolated, phosphorus emits a steady, menacing glow. *Phosphorescence* is the name that is applied to this phenomenon: it describes materials that glow without ignition. The glow of the upper ocean is phosphorescent. Some paint glows. One consistent feature of the near-death experience, reported by people whose hearts stopped beating and bodies began to fade, has been the presence of a peculiar brightness all around. Images flash, the soul floats, and the body is left behind. The mind feels calm. (It is, in fact, surging with electricity: its final moments are seemingly near.) *This is where I'm going,* people remember thinking. They see white light.

When phosphorus burns, it bonds with oxygen, creating phosphate, a molecule described by the formula PO_4: one atom phosphorus, four atoms oxygen. Phosphate is remarkably prevalent in all life-forms, although it is otherwise comparatively rare throughout the world. It is crucial to our existence. Outside of life, phosphate exists in geological form, made up of condensed, crystalline structures that are hidden in the crevices of our planet. Inside of life, it exists in every cell. It forms the membranes that hold the parts of cells together. It provides energy, in the form of adenosine triphosphate, ATP, which powers the actions of all life-forms. Even before birth, each of us gained identities by way of the cumulative influences of small phosphate groups, which held together the strands of our DNA. As we grew from zygote to cellular zillionaire, those groups enabled the replication of DNA and the formation of more complex beings—us.

The phosphorus in our bodies came, at first, from molten lava, hardened into rock. That rock eroded out of mountains, flowed down rivers, and fertilized the land below. The land supported the growth of plants, which allowed the spread of animals. The human body is, roughly speaking, 1 percent phosphorus. Phosphorus is spread throughout our cells, but it is concentrated mainly in our bones. We are extensions of the planet—we forage for phosphorus by eating plants and animals, and we fertilize the soil through waste and death. Plants thrive on this natural fertilizer. Phosphorus moves through the bodies of plants and animals, fungi and bacteria, and ultimately, usually, makes its way to the water. It is deposited as sediment: it forms new rock on the seafloor. The rock is made of compressed bodies, phosphorus squeezed from lives that are no more. It is littered with phosphatic bones, with phosphate-encrusted bivalves, with fossilized phosphate scraps. These things are hidden, set to be released in geologic time. As this time passes, the earth's plates move. The underwater rock becomes land. The land erodes. The cycle continues.

The phosphorus cycle has long connected mountains, land, and sea—but phosphate rock, pressed into geology, was mined beginning in 1847. It was used as a fertilizer, life sucked out of so much death below the ground. The growth of plants requires the creation of new cells, and each new cell needs phosphate. Phosphorus fulfills the cell's core functions; it cannot be replaced by any other element. Mined and processed, phosphate rock was applied to farm fields the world over. Crops grew like never before, injected, suddenly, with a substance that functioned like a miracle. The industry began along a sandy English cliff and grew to encompass a vast global operation.

These rocks, made of ancient lives, entered the loop in which we all once lived. Death fed life, to overabundance. Our diets changed, our lifestyles changed, and we changed. We cheated the natural process that had once governed all of our lives. Millions of tons of phosphate were uplifted every year, each ton holding bits of bodies cemented over time. The phosphorus cycle was broken in a century.

The story of phosphorus runs through every strand of DNA in every organism in the world. It runs through every piece of food and waste and every living thing. But the story of how humans changed the phosphorus cycle is rooted in a few specific spots. We first found phosphate rock in England, and the fertilizer industry began. The industry changed when rock of greater scale was found in Florida—but today, the Florida rock is almost gone. Our global agricultural system rests upon the dictates of Morocco's monarch.

Already, in some places around the world, the end of phosphate rock has occurred. It happened on the island of Nauru, far out in the Pacific, and there we see a world that passed its limits. It peaked, declined, and fell to ruin. Amid those ruins, the story of our broken phosphorus cycle comes to a close.

But it does not need to end there. There is mass resistance to the modern expansion of corporate farming methods. The world's small farmers, who produce half our food, work their land with the nuanced understanding that agriculture has always been an ecological effort. They safeguard phosphate and replenish it.

The sewage that we leave in rivers, the food scraps that we throw away, and the animal manures that run off into streams collectively contain enough phosphate to replace the amount that humans mine each year. It is possible to recycle nutrients, as small farmers still do, but the world's wealthiest nations, supported by their richest citizens—predominantly male—are leading the effort to protect the kind of agriculture that relies on phosphate mining (a kind of agriculture that, incidentally, tends to preference men who lack knowledge over women who possess it). The U.S. Farm Bill, which makes mechanical agriculture economical, is bipartisan. U.S. presidents tout corporate agricultural growth. The Rockefeller Foundation helped create the current farming system, and the Gates Foundation is working to expand it. Today, the breakage of the phosphorus cycle is worsening.

But small farmers are organizing for their rights—for land reform, gender equity, and widespread social change. Scientists, economists, and engineers are working to make phosphorus recycling compatible with modern life. Food, we now know, feeds our bodies better when it comes from healthy soils, and healthy soils come from nature, not from machines. Supported by this understanding, people are working to create a better agriculture. Cities are composting food scraps. Disenfranchised farmers are fighting for their land. If we listen to those with knowledge—rather than those with money—it is possible to restore the cycles of the earth.

There was once, long ago, a different kind of phosphate problem. When life first started, 4.5 billion years ago, the problem was that phosphorus existed only in rocks—and then, of course, no one was available to mine them. Life needed concentrated pockets of phosphorus in order to form. In a century of study, scientists have not come to an agreement about how nature solved its problem. Something happened in a pond, around a vent, near a meteor strike—something. We do not know exactly. We do know something happened, though, because we are here.

Today, phosphorus remains a part of the mix of chemical elements present in the earth's magma, and volcanic eruptions create sprawling beds of igneous rock that hold within them trace amounts of the mineral. Now, however, humanity has transferred large amounts of phosphorus onto farmland, into streams and ponds, into rivers, and, ultimately, into the ocean. The result of this is somewhat murky, but it appears that humans are changing the geology of the world. We are leaving a legacy in stone, and we are doing it by creating anew a world that once existed—one overrun with algae in the waters, with dying fish, with widespread oxygen loss in the sea. This new world is not, for us, ideal. (For algae lovers, it may be paradise.) But it is conducive to the formation of phosphate rock. This new rock will be formed and buried over intervals of millions of years. It will be hidden beneath the ground, prepared to be discovered in the future.

. . .

Just as phosphate enables life in humans, so too does it feed the life of the whale fall. The destruction of the bones of the whale provides enough fat to support a community of bacteria, and it releases enough phosphate to support the expansion of the ecosystem. The whale fall lasts because of the barrenness that surrounds it: the cold temperatures and darkness of the deep ocean preserve the whale carcass for the creatures that can access it, allowing the ecosystem to exist without floating away or being quickly eaten. Instead, whale falls remain as they begin—remote, shadowed, and teeming with life.

The nutrients provided by a whale fall represent, in a single day, two thousand years of sustenance. Their effect, ecologically, is strong enough that biologists have identified dozens of species of ocean-dwelling organisms that evolved to specialize only in whale falls, those thousands of little worlds beneath the sea. There are four-foot worms and hairy crabs, clinging shrimp and curious sharks, bacteria that float, fish that feast—a mess of life, growing and thriving, a community unto itself, separated from all other beings by a dark emptiness that extends in all directions.

This blip of abundance seems bound to recede, and eventually it will. Over a period of half a century, the whale fall's nutrients begin to dwindle, and the organisms that feasted on them go away in turn. The ecosystem fades into the landscape that surrounds it. Barrenness overtakes the ground. Just decades after a new world of opportunity opened up, life disappears; this little spot of seafloor is unlikely to be visited by such prosperity ever again.

I. LIFE

A PARTIAL MAP *of* THE EAST *of* ENGLAND

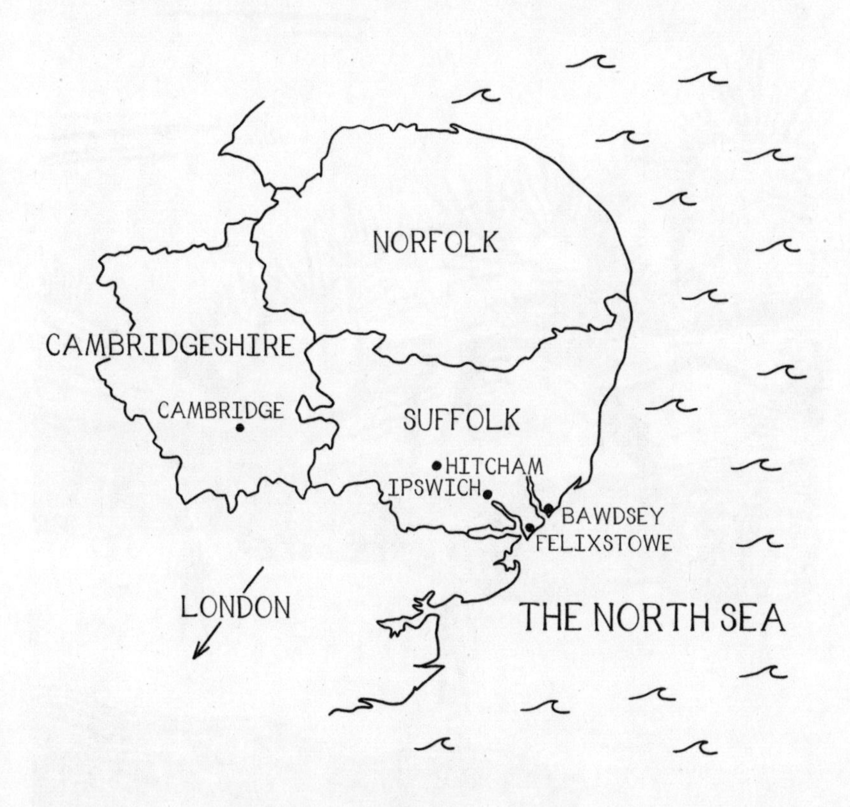

1

SEA OF FIRES

There would be no life without constant death. The bodies of our forebears came to form the base of our lives today. All of us are made of accumulated parts worn down and enlivened by billions of years of excess and limitation. Our bodies are living remnants of the past. We are all recycled—our food is dead; it was grown in soil made rich by decay. When we die, gorged and wealthy, our purpose is to fertilize the ground. The waste we create, and the waste we someday will become, exists in balance with the earth.

The world warms, and life surges. Profusions of species occupy the earth; they drink its water. They arrive in heat-driven bursts, and they flourish most in places where residues of past life can collect. Along some coastlines, upwelling waters pull nutrients from the cold seafloor. Sunlight reaches those substances on the surface, and life blooms. Elsewhere, inland seas, called into existence by warmer weather and melted glaciers, create closed circuits of growth and accumulation. Deltas, where rivers enter open water, collect the sediments that the rivers carried all along their paths. Some islands, situated far out to sea, serve as collection points for life and its excrements. In these ecosystems, life builds upon itself, and exceptional productivity ensues.

Something changes on the earth. The temperature cools, the continents shift, the oceans drop their nutrients somewhere else. The closed circuit opens, and bodies are covered in sediment. They settle and condense, growing firmer under pressure from above. Chemicals dance—crystals form. Metals dissolved in seawater are absorbed by what remains, and the crystals fit more cleanly with one another. They are stable, and then they change again. Storms and tides churn the seafloor, washing away some materials and bringing others near. Years pass, and the continents move again, pushing compact crystals out of the water. Settled and churned, scarce and hidden, buried beneath layers of geology, this collection of reworked matter is a literal crystallization of life. It exists in a thousand different places and a thousand different forms, holding contents of silent value. This is phosphate rock.

Phosphate, underneath the surface, holds the power of life itself. That power is exerted through the forces that shape the physical world. All of life is controlled by geology. Through the steady, cyclical accumulation and release of phosphate, unseen forces determine the biological productivity of the earth. When the rate of erosion increases—when more phosphate is added to the system—the world is overrun by life. When erosion decreases, or a large expanse of phosphate is deposited as rock, the system loses phosphate, and life's sum total drops. Phosphate exists within the confines of a natural order, a deliberate cycle of destruction and creation. It travels slowly between life and rock, and its journey defines the contours of the life that can exist on earth. It is an adherence to the plodding of geologic time, but with a twist: some hidden crystals, in recent years, have begun to travel at a different pace, in a different direction, along a different path. This new path began along the coast of England, at a time when the countryside—the sloping plain of Suffolk in particular—was on fire.

The arsonists were everywhere. They were living in the villages and working in the fields. They came out in the dark of night and crept onto the lands of wealthy farmers and lit up their barns and sheds, their threshers and carts, their haystacks. They killed their animals. They were rageful—and they were poor.

England's wealth was growing as the dignity of the poor declined. Land was privatized, child labor thrived, and the poorhouse system punished society's weakest members. In East Anglia, the region that includes Norfolk and Suffolk, the situation was particularly dire. The textile economy had declined in response to London's growing industry, and the workers were left reliant on agriculture. But farmwork was seasonal and unsteady, and it was controlled by landowners and by farmers, the powerful occupiers of the land. Population grew, wages declined, the poor grew poorer, and the land became weaker. Then came 1843. A dramatic hailstorm destroyed the crops in August, and a terrible drought arrived the following year. Little grew, and the laborers could not work. People could not pay for food; there were hunger riots. And as they starved, they watched shipments of food going elsewhere, although the landowners continued to make a profit from their crops.

In 1843 and 1844, more than six hundred fires were set across England. Arson attacks were carried out against unjust magistrates and landowners who evicted tenants. Fires were set at workhouses, at jails, at taverns, and at shops. By far the most common targets were the farmers whose decisions had consigned the laboring classes to desperation and shortage. Entire properties were destroyed. "The year 1844 was one of intense alarm to the owners and occupiers of farms," the historian John Glyde wrote in 1856. "Fires were of almost nightly occurrence during the long evenings. The occupiers of land lived in a state of nervous excitement, looking about their premises every night before retiring to rest, apprehensive of their crops being destroyed by the match of the prowling incendiary." "Even among some of the poor the fear took on almost apocalyptic proportions. Two preachers, Winter and Burgis, tramped the countryside warning the people the world was about to end in one great catastrophic fire," the historian John Archer wrote. "The end of the world was fixed for 21 March 1844."

At the center of the fires was the village of Hitcham, and at the height of the flames Hitcham was led by a famous man. The naturalist John Stevens Henslow, appointed rector seven years before the world's forecasted end, was an academic of national renown who worked as a professor at the University of Cambridge. He was first a botanist and then a mineralogist, a researcher whose focus was on the natural world. When Charles Darwin came to Cambridge in 1828, Henslow had taught him botany. At the time, Darwin was struggling. He felt disconnected from his studies and had trouble keeping up at Cambridge. During those years, Henslow and Darwin embarked upon a series of adventures that inspired Darwin with a sense of the boundlessness of the natural world. The two went searching for flowers in the countryside and traversed the swampy fenland north of the city. They made aborted plans to visit the Canary Islands. Henslow saw in Darwin early traces of the mind that would revolutionize scientific thought. When the HMS *Beagle* was launched in 1820, it was Henslow who ensured that Darwin was aboard—although Henslow himself had received the initial invitation to go. While the *Beagle* was away, Darwin and Henslow communicated extensively, and Darwin had his trove of specimens, collected from around the world, delivered to Henslow in England for safekeeping. Henslow, Darwin said, had changed his life.

The same year Darwin published *The Voyage of the Beagle,* Henslow moved to Hitcham for good. It may have seemed a surprising change, but the 1830s were a difficult time for the university, and a government-funded rectorship at Hitcham paid far more than a Cambridge professorship did. Henslow scaled back his academic duties, and by 1839 he was living in the countryside and running the parish. The conditions he found there shocked him. Hitcham had long been a place of discontent and disruption. In 1820, the parsonage had been set aflame in response to a denial of poor assistance for a sizable number of residents. Seven years into his rectorship, the harvest had failed, and the violence had worsened. "We have almost nightly fires about the neighborhood," Henslow wrote in 1844, "and as I was lecturing at Hadleigh on Wednesday, a cry of fire interrupted and spoilt all." In April 1844, a correspondent from *The Times* of London reported an epidemic of nightly crime in Hitcham, a scourge that included frequent fires and a number of break-ins at the church, the corn mill, and many shops. "Scarce a night passed during the winter months, at one period, but the sky was lit up with incendiary fires," Leonard Jenyns, Henslow's brother-in-law, would write. "No farmer felt himself secure."

The crux of the problem in the countryside lay in the uneven production and distribution of food. In a decade of want, during a century of growth and unrest, the prospect of food shortage loomed large in the minds of Henslow's parishioners. Some of this had to do with what was happening across the Irish Sea. A potato blight first appeared in 1845, brought over from America, and quickly spread through Ireland, where widespread famine took hold. Even though the Irish grew a wide array of crops, the potato was the staple of the diets of the poor. The rich maintained a profitable export trade of other crops, and the government declined to intervene. A million people died of hunger. The threat of famine in England presented similar concerns. Facing down the landed interests of Hitcham—the landowners and the farmers—Henslow used his role as rector to press for social change, sermonizing about Christian charity and making overt appeals for financial assistance for his schemes for the poor. He advocated for higher-quality education and for more recreational opportunities for the working class, and he founded a school, funded through his own salary, for the children of laborers. He spoke openly of the need for farmers to respect the rights of workers—a kindness but also a necessity as those abused workers burned the countryside in despair.

Henslow felt that science and social well-being were inextricably intertwined; curiosity and scientific investigation, he believed, could improve the lives of common people. An influx of agricultural science would solve food shortages, quelling unrest. He fashioned a series of experiments that involved dozens of farmers from around the parish working to raise yields through the use of natural fertilizers. He lectured on farming methods and soil fertility and published fifteen Letters to the Farmers of Suffolk in a local newspaper. He gave lessons on manuring, taught botany to children, and disseminated scientific information among his parishioners, many of whom were illiterate. For those who could read or were learning to do so, he founded a lending library, instituting adult literacy classes shortly after. "The *art* of husbandry should be converted into the *science* of agriculture," Henslow was known to say.

A scientist walking through his parish, Henslow was often seen investigating the rocks that made the roadways or the soil that made the fields. Rector and also magistrate, he remained, over time, a naturalist. Like many scientists of his day, he had received broad but scattered training in various natural fields. He had been drawn to science from a young age, exploding gunpowder, modeling caterpillars, and collecting fungus and flowers at home. At school in London, he was entranced by the idea of travel. At Cambridge, as an undergraduate, he studied mathematics. He participated in a geological survey of the Welsh island of Anglesey, cataloged a vast collection of British plants, and studied the formation of mineral crystals. Among his most lasting contributions was the Cambridge University Botanic Garden, which he managed and moved to a new location, nearly eight times larger, devotedly collecting specimens that made the garden a botanical masterpiece.

In 1842, during a holiday in Felixstowe, Henslow took a walk along the North Sea coast of Suffolk, and he saw stones and fossils within layers of clay, all of it buffeted by waves. Among those pieces, he found a rock that would launch the modern phosphate industry.

One hundred and seventy-nine years after Henslow's walk, Neil Davies, a sedimentary geologist at the University of Cambridge, is standing at the base of a cliff layered with red sand, the North Sea at his back, holding up a pound's worth of bulbous, dark material that might be phosphate. "It could have been some kind of nodule at the time—it's got that kind of cracking inside it. Unlike the stuff in the Greensand and the Gault, you don't often get to see what these things are, just because, after they were deposited, once they were fossilized, they were dragged out and rolled back and forth repeatedly by waves." Davies and I are searching for phosphate in the cliff—the Bawdsey Cliff—where the modern phosphate industry began, where the wastes of life and the promises of death lost their balance with each other, where humanity upended the limits of nature and launched itself into a period of interminable growth.

We are here to see the Red Crag, a layer of sandy strata that formed 2 million years ago. The Red Crag sits directly upon the London Clay. The Clay—thick, blue gray—formed between 56 and 49 million years ago, when eastern England was submerged in a vast, warm, life-enriching sea. We are standing on the London Clay, looking at the Red Crag; the border between the two layers lies roughly at our knees. "If you were to take out that block of clay there and wash it down, you would find a handful of pieces of phosphate, a handful of nice fossils inside there. But because they're so randomly scattered throughout it, and because the clay is so resistant, it's not like you see them sticking out very often," Davies says, as I reach to break off a chunk of damp material. "Clay is a cohesive sediment. The particles are so small that they all are chemically attracted to one another—plus, of course, getting wet, they basically become one agglomeration of particles, whereas the sand is sand: it just falls apart in your hands as you touch it."

What is known is this: the London Clay Sea was an expansive body of water that once, long ago, extended from the middle of England to the plains of Poland and north to Scandinavia. Covering much of northern Europe, the sea was isolated, sharing characteristics with the Mediterranean Sea of today. In a series of successive waves of sedimentation that occurred over a period of 7 million years, the London Clay Sea deposited a layer of material that is, in some places, five hundred feet thick. Because of the features of the sea—the warmth, the productivity, the occasionally oxygenated bottom—fossils were preserved, and phosphate formed. The clay preserved the life that had inhabited the surface for millions of years.

There is a gap in the sedimentary record that spans the distance between the end of the London Clay and the beginning of the Crag, an interval of time equal to about 45 million years. The lack of sediments suggests that, during this time, the sea receded, and the top of the London Clay became dry land. It stayed that way until a little over 5 million years ago. At that time, the earth warmed, and the sea came back over the land. The resulting series of resurgences of water created a sedimentary layer collectively known as the Crag Group. Four sedimentary deposits were made over a 4 million year period: Coralline, Red, Wroxham, and Norwich. As they stand today, each deposit represents, at most, a couple hundred feets' thickness of collected sediments. Torn away by glaciers and weathering, the deposits are not continuous layers across the surface. They are an occasionally overlapping checkerboard of geological history, spread across the North Sea, that slides up out of the water to blanket more than two thousand square miles of eastern England. The quality they have that interests us is that, in growing, they reworked the London Clay. As the Crag seas expanded and receded, they churned up strata below. They released the fossils and phosphate from within the London Clay. Like so many little incipient whale falls, those materials floated through the Crag seas, they moved with the tides, and they joined new geological layers. Generally, they settled in lags along the bottom layer of the Crag, resting just on top of the London Clay, along the boundary visible at Bawdsey, before us, at our knees. They sat there for eternity.

The drive from Cambridge to the coast takes an hour and a half. To get to the Bawdsey Cliff, Davies and I—along with William McMahon, a postdoctoral research geologist at Cambridge—had walked around a spit of land that pierces the river Deben across from Felixstowe, a town whose port is the largest in England. We climbed along a narrowed bit of silty beach, crossed a field of sea cabbage, and moved along an embankment made of iron. It was a walk Davies has made many times, and he narrated landmarks along the way. There was Bawdsey Manor ("an early radar experimental station"), Sealand ("this little rogue microstate"), and Pulhamite ("a good material for tricking non-geologists"). Bawdsey is the clearest place anywhere to see the border between the London Clay and the Red Crag. The difference there is as clear as day. Several feet up the cliff—if the sea hasn't shifted the beach, or trees haven't washed up sideways, or erosion hasn't hidden the unconformity—the blue stops and the red begins. The distance between the layers, a space traversed by phosphate nodules during reworking, is nearly 50 million years. The Red

Crag, in this location, rises about ten meters into the air. Below the ground, the depth of the London Clay is likely many times that.

The research that Davies and McMahon conduct is generally focused on the relationship between life-forms and sedimentary rock. The geologists are curious about the ways in which organisms affect the rock record. One of the key field locations for their research is the Torridonian Group, a formation in northwest Scotland that is 1.6 billion years old. Contained within the Torridonian are rivers that existed before plant life proliferated—rivers, with paths unaffected by grasses or trees, that widened and moved in patterns different from those of the present day. Without plants or animals, samples of stratigraphy become much harder to date. Time turns into a more ambiguous construct. There are fewer telltale fossils left behind, fewer time-stamped burrows or peculiarly aged organisms to use to narrow down a range of dates. Their age can be a mystery—or, at least, a basis for prolonged attempts at understanding. Davies, who earned his PhD in 2003, has carried out research in twenty-one countries. His interest in phosphate stems from questions about the relationship between life and the creation of rocks, and between rocks and the creation of life. If life gathered phosphate, geologists ask, from whence did phosphate come?

The layers of the earth are shaped by the waves of life that cover it. Limestone, for example, is predominantly created by the compaction of the shells of water-dwelling organisms, including corals, oysters, and clams. Calcium carbonate, the main ingredient in the shells of those creatures, is, by extension, the main ingredient in limestone. These organisms filter calcium out of seawater, using it for self-construction. When they die, their shells are left behind and slowly broken down, pressed together, and regenerated into rock. The presence of limestone today is an indicator that an ecosystem thrived long ago.

Chalk, like limestone, is a kind of rock that suggests past life. White and crumbly, chalk is formed of algae collected and crushed over an interval of millions of years. The algae, in order to exist, required warm temperatures and an adequate supply of nutrients and sunlight. The white cliffs of Dover descend from algae that grew in the Cretaceous. The algae solidified on the seafloor and came to the surface as the African and Eurasian Plates collided. Of those dramatic histories, only chalk outcrops remain. "Phosphate is one other clue about what's happening in the earth system. It's telling us one thing about what organisms are using to build their skeletons, what kind of nutrient cycling is going on in the ocean," Neil Davies said. The formation of phosphate rock suggests a time, long past, when phosphorus was in excess—when life surged and was extinguished. During times of phosphate rock formation, organisms grew so much that oxygen periodically disappeared, a necessary shift for the stability of the new mineral. At Bawdsey, the environment was warmer than it is today, but it experienced periods of deprivation. "The phosphate was produced back in the Mesozoic and the early Cenozoic. And it became phosphatic rock. And then, sometime later on, two million years ago, there were waves beating against that rock that liberated it, brought it out, deposited it into these Crag seas, and for a moment, from those outcrops at the present day, you can see that snippet of that little aspect of history."

The Red Crag crumbles, and the London Clay cracks. The Crag is dry and dusty and tumbles groundward at the slightest touch, the fossils, rocks, and shells it holds flowing freely down an embankment of sand. The cliff is a wall of loose particles occasionally organized into discrete layers like a stairstepped chain of cantilevered shelves. When touched, the layers fall apart, disintegrate, and reveal, sometimes, a new fossil—perhaps a piece of phosphate. When the sand dribbles down the side of the cliff, it passes out of the Crag and over the Clay, covering that blue-gray compacted mush with a smattering of burnt-orange dryness. The roots that burrowed downward from above are left dangling in the air. The red sand reaches the sea.

From a distance, the London Clay below the Crag appears somewhat like a hard igneous rock—a rough mountain outcrop, set by heat and worn down slowly by erosion. Up close that image dissipates, and the dark clay squashes at the touch of even a gentle, curious hand. If a large chunk is pulled off—a piece, say, the size of a fist—and squeezed, it breaks in half. If those halves are squeezed, they split, too. The particles stick to one another until the tiniest constituent part of the block is pressed between a finger and a thumb. Even then, it hangs together, forming more a sheet than a dust. There are objects in the clay—fossils, shells, phosphate—but they remain stubbornly locked in a material that does not set them free. In contrast to the Crag, which is often only about ten feet thick or washed away completely, the Clay extends hundreds of feet underground.

In the Red Crag Sea, deposits came in bursts. Tidal currents set down inches—often feet—of sediment over spans of days or weeks. The churning seafloor built up randomly. The sedimentary record consists of an exploding network of pockets of rapid deposition. A geologist, observing a seam of phosphate in the Red Crag, can reasonably infer that it was laid down over an interval of about a month. The deposit as a whole took 2 million years to complete, and it happened by way of bits of seams settling haphazardly throughout the ancient North Sea, out of which over 2,500 square miles of seafloor are now perched onshore in Britain. A single seam of reworked phosphate fragments may have taken a month to collect and settle into place, but a nearby seam, some number of feet away, may have formed many thousands of years later. At a higher level, the process was gradual—but, close-up, it was imbued with violence. There was a suddenness to its creation. When these nodules were deposited—50 million years ago, then 3—they took phosphate from the bodies of fish and other organisms, and they left it in the record of the rock.

The rocks that Henslow found at Bawdsey were nodules that had formed in such a way. They were smooth, dark, fist-sized. Some of them were round, others curved or straight. They were packed with fossils. They were mysterious, and Henslow returned to Bawdsey in 1843, bringing along his son George, to investigate again. In 1844, he brought samples to a meeting of the Geological Society of London, opining on their origin before the august members of the group. He wondered whether they were coprolites: spiraling, phosphatic bits of dinosaur dung.

A year later, writing for the British Association for the Advancement of Science, Henslow put his discovery into perspective. "The supply of phosphate of lime used in agriculture, and hitherto obtained from bones, having of late years become insufficient," he wrote, new sources of material would soon need to be found. "In October,1843, Prof. Henslow"—referring to oneself in the third person was the accepted style of the publication—"had called attention to the occurrence of phosphate of lime in pebbly beds of the Red Crag at Felixstow [*sic*], in Suffolk; these nodules, though extremely hard, presented external indications of an animal origin, and yielded, upon analysis, 56 per cent. of phosphate of lime." This was an astoundingly high number: the accepted minimum value for a nodule to be considered phosphatic today is about 30 percent. The rocks that Henslow had found on a whim were made up more of phosphate than of all their other components put together. Usually, the challenge of phosphorus is that it is overly spread out; Henslow had discovered a place where it was concentrated, and to an extraordinarily high degree.

In an 1845 letter to *The Gardeners' Chronicle,* Henslow explained his discovery to a wider audience. "It was in the summer of 1842 that my attention was directed to certain nodules in the red crag, between Felixstow [*sic*] and Bawdsey," he wrote in the magazine. "In the summer of 1843, I paid closer attention to these nodules, and, after inspecting a very considerable number of them, I arrived at the conclusion that their origin was somehow or other connected with the decomposition of animal matter."

The mining, when it began, consisted of small teams of diggers carving holes in the coastline around Bawdsey. It was 1847, five years after Henslow found the rocks. "The opening of a new pit was no small job in the days when every thing was done by hand," the journalist Walter Tye wrote for the *Suffolk Chronicle and Mercury* in 1949. "So for days and days they had to dig and dig until they struck the seam. Whilst small pockets were sometimes found near the surface, they usually had to dig down from thirty to forty feet before striking the main seam." Systems of scaffolding were constructed inside sloping pits, with men and boys working on the sides. Wheelbarrows were pushed along narrow planks to move the phosphate from the bottom to the top, where it was sorted, carted to big sluices of water to be washed, and loaded on barges to be sent upriver and processed into fertilizer. Generally, the digging was done by hired farmworkers. Between 1847 and 1889, somewhere in the range of ten thousand to twelve thousand tons were mined annually in Suffolk, mainly from areas near the river Deben. Processed into fertilizer, the mined rock was shipped to farmers to increase yields. "Veins and pockets were found on most farms in the district, and as much as £20 worth were often dug from a cottager's garden," Tye wrote. "No able-bodied man in the locality had any fear of unemployment, for jobs could always be found in the pits."

In a 2009 paper, the local historian Bernard O'Connor and the University of Leicester geologist Trevor Ford estimated that 325,000 tons of phosphate rock were mined in Suffolk between 1847 and 1874. "The original workings were on the coast at the foot of the cliffs at Felixstowe. There'd been a landslip which exposed a bed of fossils. . . . Having removed the fallen overburden, you'd have to dig out any exposed fossils. That entailed undercutting the cliff, which then collapsed. The landowners on the top of the cliff weren't keen to lose their property on the edge, so restrictions were imposed on the undercutting of the cliff. The mining then moved inland up the river Deben. And where the river had cut through the fossil deposit, it was possible to see the bed of fossils exposed on the side of the riverbank. So then they would start digging into the riverbank, taking the fossils and putting them into wheelbarrows, perhaps putting them into sacks or just putting them into the barge," O'Connor, who spent much of the 1980s studying the history of phosphate extraction in East Anglia, told me. "That's how the mining expanded."

The mining expanded, and an industry was born. Agricultural laborers, jobless, went to work as miners and producers of fertilizer. Stability returned to Suffolk and the rest of East Anglia. For the rest of the world, the changes proved consequential, although few outside the scientific community noticed at the time. The new phosphate industry made millionaires of local landowners, but its scale was small compared with the enormous requirements of the farmers of the time. It was only the beginning of something big. Stores of material that had been hidden would soon be freed by a global industry. That material would become fertilizer and then food. It would enter the bodies of billions of people. It would change agriculture—but first, more discoveries needed to be made. Scientists needed to develop a better way of making phosphate fertilizer, and the rocks at Bawdsey would provide the stimulus for that. Henslow's stones would enter first the laboratories of scientists and then the factories of businessmen, and the relationship between humans and the earth would change forever.

Henslow understood the importance of what he found. Yet, after making his discovery, he soon directed his attention to other endeavors. He did not participate in any kind of mining. Instead, he devoted himself to his village and his scientific work. Darwin continued to come to him for advice. Queen Victoria did, too: Henslow was invited to Buckingham Palace to teach botany to the children of the royal family. He based his lectures on the same curriculum he had taught to children in Hitcham. In 1850, Henslow became president of the Ipswich Museum, a post he held while continuing to support the Cambridge University Botanic Garden as well as the institution that became the Royal Botanic Gardens at Kew. He organized the instruction of botany at the newly formed University of London, a public institution that realized his dreams of providing education to a broader portion of the population. At Cambridge, he helped to establish the Natural Sciences Tripos, an interdisciplinary framework that governs instruction in the sciences to this day.

But by 1861, Henslow's body was failing him, and in May that year, he died. "The paroxysms occasioned by inability to breathe freely were, it is said by those who waited upon him, as painful to witness, as they were agonizing to himself to bear," Leonard Jenyns, his brother-in-law and biographer, wrote of Henslow's final months. "His great natural strength gave way very slowly, and resisted, beyond all the first expectations of his friends, the encroachment of his disease. It was, to use his own expression, 'like granite wearing away from drops of water.'" A simple ceremony was conducted at the Hitcham rectory, and then Henslow was laid to rest. As relatives took his body from the church, villagers gathered outside to say goodbye. The rector of Hitcham was gone.

The countryside was quiet. After the waves of chaos of the 1840s passed through England, such violence had dissipated. In Hitcham, Henslow had implemented his reforms, turning the village into a progressive beacon for the county. The results were palpable. Phosphate was more plentiful, and farmers better understood the process of adding fertilizer to the land. Given food and education, science and self-respect, the parishioners of Hitcham had settled down. They were, wrote John Archer, "the very models of virtuous rusticity."

Five years after Henslow died, Leonard Jenyns spoke about phosphate to the Bath Natural History and Antiquarian Field Club in England's southwest. "As we look at these shapeless stones, they seem to have a voice which leads our thoughts above the world to its One great Ruler—even the stones cry out and speak to us how through countless ages He prepared the world for man—His greatest and last creation—how for man He hid beneath the surface of the earth treasure houses of precious things which from time to time he brings forth for their welfare and provision. Yes in these little stones let us see the tracings of the fingers of God," he said. In the remnants of ancient death—in the ashes of the fires of past community—new life had been reborn. The bodies of our forebears, the accumulated pieces of biology, the weathered leavings of worldly life, had finally been uncovered. Beneath the surface of the earth, the treasure of creation had been found.

2

THE ACID TEST

The excrements of animals vary in the amount of phosphate that they hold. Per pound, the phosphate contained in bird droppings is often twenty times that found in manure from cattle. Pig feces lies somewhere between the two. The nutrients found in bird waste attract plants, which form tiny ecosystems around it. Within one such ecosystem exists the Surinam cockroach, which itself has adapted to thrive among the plants that grow surrounding bits of waste. The cockroach takes the nutrients and transfers them to a different form. It moves them across the landscape. When the cockroach dies, its body returns to the ground. Its phosphate, once again, becomes available. Another organism grows.

This is, in essence, the phosphorus cycle. It is the movement of atoms between biology and geology, between water, land, and life. It is an interaction of continual loops, a complicated interchange in which every living being plays a part. The phosphorus contained in rocks is broken down by weathering, by wind and rain and environmental acids, but also through the slow, forceful attacks of mosslike lichens, which release crystallized phosphate for the rest of the world to use. Lichens, in their entirety—gray-green, growing, paper-like, on boulders—are organisms that exist in furtherance of the phosphorus cycle. The force of their burrowing is greater than the power of a car. It is sheer violence, biological creative destruction, a continual reduction of life to its basic ingredients. As a lichen grows and dies and rots, it produces phosphorus. Plants absorb that phosphorus, providing it for animals and other organisms to consume. When a deer rummages around a forest, eating grasses, berries, and seeds, it is collecting phosphate and releasing it elsewhere, by way of dung. When a dung beetle perches atop a piece of feces, it is completing the transaction.

The process of erosion is a slow enrichment of life downstream. When the Indian subcontinent crashed into Asia and formed the Himalayas 50 million years ago, the new mountains underwent intense weathering that released massive amounts of minerals into the oceans, leading to a surge in life that continues today: although that range represents only 4 percent of the world's drainage area, it accounts for one-quarter of the nutrients that flow from rivers into seas. The Himalayas, through intensive weathering caused by rapid tectonic uplift and heavy rain, feed the world's oceans. The elements they bring up from the earth are nutrients that contribute to ecosystems. The life-forms of the sea owe some of their abundance to the growth of the Himalaya Mountains. The global fishing industry benefits from the geologic speed of the Indian subcontinent. Humanity gains when mountains crumble.

Salmon migrate upstream, carrying phosphate with them. A bear eats a fish and then leaves its waste upon the land. Beavers dam streams, causing wastes to collect. In the ocean, schools of fish move up and down like clockwork every day, following the sunlight and opening a new wealth of nutrients for predators to consume; it is an enormous, global transfer of nutrients that never ceases. Dead plankton sink into the darkness below, fertilizing life. A dead whale drops beneath the waves, doing the same. On land, around the world, most of the food eaten by elephants goes undigested and unused. It tumbles through the elephant's body, broken down but unabsorbed. The effect of this occurrence is that nutrients, held high up in the trees, are lowered to the surface of the earth. They are consumed by plants and animals. The cycle presses on.

Phosphorus is the element of life. Phosphate is the mineral of the life-adjacent. In 1842, as Henslow picked up pieces of phosphate along the coast, other scientists were looking for a better way of making phosphate fertilizer. Using those new supplies of phosphate rock, researchers in England and Germany created fertilizers, and businessmen near Cambridge worked to sell those fertilizers to the world. New companies were reducing life to its chemical bits and then repackaging it. These companies—which led the way for the chemical giants of today—changed the way humans participated in the phosphorus cycle. The change was significant, but it had a precedent: in many different ways, in many different places, the phosphorus cycle has long been affected by the presence of humans on earth.

Although John Stevens Henslow first found phosphate in 1842, farmers in Suffolk had long made use of phosphate nodules found in sand pits on their properties. There were rocks within the ground outside of Cambridge that were known to make plants flourish. Phosphate was a part of local lore. "Most of those farms in East Anglia, around Ipswich, had their own little Crag pits. And they were extracting phosphate to put on the fields as fertilizer," Liz Harper, a paleobiologist at the University of Cambridge, told me. She grew up in Suffolk. "I've seen pictures, goodness knows where, of pits in farms with literally just holes dug in the ground. And a bloke standing there with a big wooden pole stirring sulfuric acid with these nodules in a very, very primitive way of extracting the phosphate." Henslow, scooping up rocks at Bawdsey, had only uncovered what the locals already understood.

The life Harper studies bridges eras of time. She researches mollusks—a diverse group of invertebrate animals that includes octopuses, snails, and squids—and she focuses on bivalves, which are, generally speaking, mollusks of the shelled sort, including oysters, mussels, and clams. These are ancient organisms that first entered the fossil record during the Cambrian explosion half a billion years ago. Many species have changed very little since then, so Harper studies both prehistoric fossils and modern specimens. The easiest way to learn about an ancient organism, sometimes, is to study a modern one. Bivalves have persisted all these years because they have evolved, with great precision, to carry out a single task: water filtration. They are constantly sucking in liquid and removing its nutrients. Slowly, over the course of eons, they are taking phosphate from the water and leaving something edible in its place.

Filter feeders are the foundation of the food chain. Their purpose is to make nutrients usable for other organisms. Once they remove phosphate—as well as nitrate and a number of other compounds—from the water, they turn it into flesh that can be consumed. More bivalves mean more flesh, and it means less phosphate in the water. A loss of bivalves can cause the chemistry of an ecosystem to change: they regulate the phosphorus cycle. In Virginia, they are held up as a solution to nutrient pollution, and communities are urged to grow oysters in an effort to absorb the excesses of a fertilizer-polluted ecosystem. In Lakes Michigan, Huron, Erie, and Ontario, thanks to the introduction of new species of bivalves, which some biologists have referred to as an invasion, the problem is the reverse. A 2021 study found that "phosphorus cycling in the invaded Great Lakes is now regulated by the population dynamics of a single benthic species, the quagga mussel." Bivalves, at the time of the study, constituted the overwhelming majority of nonplant biomass in the lakes. The water was devoid of phosphorus. The chemistry had changed.

Like all organisms, people, by our very nature, affect the cycles of the earth. The redistribution of phosphorus by humans began long before the first creation of phosphate fertilizer in England. It began even before the birth of modern agriculture. Just as salmon carry phosphate upstream and land animals concentrate it, people have collected and moved phosphate since the birth of the species. We drop it in our excrements, gather it in food scraps, and leave it in our bodies when we die. The difference between foraging and farming societies is sometimes unclear. The camp-follower hypothesis suggests that agriculture came to the fore because weedy plants—which prefer disrupted landscapes—grew in the wake of hunter-gatherer settlements. The phosphate-heavy wastes that were deposited around these settlements made the soil rich, encouraging plants to reproduce. They evolved, people returned seasonally, and the plants were slowly domesticated. In 1993, based on experiments conducted in the Asana valley of Peru, the anthropologist Lawrence Kuznar suggested that quinoa, the Andean grain, developed symbiotically with foragers in modern-day Bolivia and Peru. The waste generated by humans and the animals they herded was more concentrated than what would otherwise have been the case. Llamas, raised by humans, ate wild plants of the *Chenopodium* genus—from which quinoa descended—and left seeds in these highly fertilized areas of land. The plants, unsurprisingly, thrived. They evolved; they domesticated. They became quinoa. According to Kuznar, modern quinoa arose from the patterns of prehistoric human waste.

Researchers of early humanity have found extensive evidence that people living in preagricultural societies engaged in practices that, at the very least, resemble farming. Many of these techniques relied on fertilizers. Hunter-gatherers, moving in groups, concentrated waste in the landscapes they inhabited. They shaped the ground by fertilizing specific portions of it. They created networks of rich soil. The historian Charles C. Mann has written about how European explorers, arriving in Australia and the Americas for the first time, found continents that had already been altered by human habitation: they described parklike landscapes, grazing animals, and concentrations of edible plants that could only have resulted from the deliberate efforts of humans. The Amazon rainforest, held up today as an example of untouched wilderness, appears to have been partly or largely cultivated. Patches of soil, referred to by archeologists as Amazonian Dark Earth, were apparently maintained in order to create rich growing conditions within the otherwise poor soil of the biome. Organic materials—human excrements but also animal bones and dead fish—were systematically added to the ground alongside charcoal, which, at a chemical level, held the nutrients in place. The soils, in all, were preserved and slowly enriched. Those patches of land, also known as *terra preta,* allowed generations of agriculturists to prosper, redefining the rainforest's ecology. The vast quantities of fruit trees spread throughout the Amazon, too, are no coincidence: in controlled experiments, anthropologists have found that characteristics of the Amazon's supposedly natural ecosystem—the uncanny prevalence of edible foods, particularly fruit trees—occur only in human-managed areas of land. Estimates of the portion of the rainforest that was cultivated by indigenous human inhabitants range as high as 90 percent. ("Visitors are always amazed that you can walk in the forest here and constantly pick fruit from trees," the anthropological botanist Charles R. Clement says in Mann's 2005 book *1491*. "That's because people *planted* them. They're walking through old orchards.")

The rise of full-time farming changed humans' relationship with the land, making soil a commodity to be preserved the world over. Waste became an important material: it enabled investments in the ground. Fertilizers were used by early agriculturists both to safeguard rich farmland and to begin cultivation of soils that started out poor. In Senegambia, ancient rice farming techniques called for cattle to be grazed on cropland just before planting every year, in order that their manure might fertilize the fields. Pre-Inca civilizations in South America, including the Paracas and the Moche, made poor soils suitable for agriculture by hauling bird excrements, called guano, from outlying islands and applying them seasonally to their crops. These accumulated excrements possessed high levels of phosphate and nitrate—levels much higher than those found in conventional manures—and allowed small empires to flourish in the otherwise arid Atacama Desert. Beginning around the year 1000, the Inca expanded to become the largest empire in the world. The borders of that empire, which stretched over two thousand miles along most of the western coast of South America, matched precisely the extents of the habitats of nesting seabirds. The guano trade—and the conservation of the birds—became a central component of Inca policy and practice. Despite encompassing vast regions of mainly poor soil, the Inca managed to maintain consistent food surpluses every year.

Ancient peoples learned to use the cycles of nature to their advantage. Early civilizations along the Indus River benefited from annual cycles of flooding that pulled phosphate-rich sediments from high in the Himalayas downstream, where they fertilized the river's floodplain. The strength of early Egyptian civilization rested largely on the fertilizing power of the Nile River, which swept over the land each year and stayed for more than a season, depositing nutrient-rich sediments that kept farmland healthy. The Nile, Herodotus wrote in the fifth century BCE, "rises of itself, waters the fields, and then sinks back again." Suspended in the water were sediments carried from various points along the river's four-thousand-mile length. Notably, the Nile's headwaters incorporate the Ethiopian Plateau, a highland area made up of magma hardened and left for 75 million years. The flood basalts that undergird the plateau are rich in phosphate. The rains that bring the Nile's floods erode the rocks of the plateau, pulling phosphate down into the currents. At the time Herodotus visited Egypt, annual summer floods caused the river to grow many times its normal size, engulfing a wide swath of farmland. Farmers designed complicated systems of canals and dikes that maintained high water levels even as the river began to fall again, giving the sediments time to sink into the soil. The black silt that coated the fields contained the Ethiopian phosphates, which enabled the growth of crop plants in the months that followed the flood. Herodotus, observing farmers in the Nile delta, marveled, "There are no people, either in the rest of Egypt or in the whole world, who live from the soil with so little labor."

On the other side of the Mediterranean, where water came from rain more than from rivers, greater effort was required to maintain the fertility of the soil. Theophrastus ranked manures—he placed humans' at the top—while Columella, the Roman agriculturist, advocated compost piles for gardeners who did not have access to animal waste. "See that you have a large dunghill," advised Cato the Elder in *De agri cultura,* the oldest known work of Latin prose. (In Latin, as it happens, the word *laetamen,* for manure, derives from *laeta,* which is used to describe lushness. The goddess Laetitia is often shown with flowers.) In addition to manure, fertile marl and a selection of green plants including clover and lupines were plowed into the soil. In the second of Homer's epics, Odysseus returns home to find his dog sleeping "on piles of dung from mules and cattle, heaps collecting out before the gates till Odysseus' serving-men could cart it off to manure the king's estates." In Europe as a whole, manuring had been practiced for six thousand years by the time Cato recommended its use to the Romans. The period of time during which manure has been applied to the land in Europe encompasses all of agricultural history on the continent: farmers, from the outset, fertilized their crops.

The early riverine societies used those methods, too. Aryans living near the Ganges four thousand years ago standardized the application of manure as part of a broader shift toward agricultural technology that included weather reporting, the use of horses, and the construction of dams and canals. The Egyptians viewed dung beetles—those vital decomposers of excreta—as sacred, placing them in tombs. They also burned weeds and ground phosphate-rich seashells to augment the fertilizing effects of the Nile. In Mesopotamia, the act of composting paralleled the development of agriculture from ten thousand years ago to the present. As the Tigris and Euphrates made deserts arable—at least in the short term—ancient Mesopotamians collected excrements and used them to grow rich fields of grain. In those fields, which bordered population centers, broken ceramics indicate the painstaking movement of manure from city to farmland. Across these societies, the combination of water and fertilizer prompted agricultural surplus that spurred political might. Somewhere within the Mesopotamian landscape, the Hanging Gardens of Babylon, which may or may not have existed, were reputed to raise plants in a hydroponic fashion, with fertilizers fed to them in aqueous solution.

Ancient systems of fertilization were not limited to individual endeavor. Five thousand years ago, the ancient Minoan civilization built a sewage system that reused wastewater as fertilizer. Two thousand years later, ancient Athenians did the same thing, sending their excrements to Cephissus, outside the city, to fertilize olive groves. The geographer Ellen Churchill Semple points to countless biblical examples—from Exodus, Isaiah, 1 Samuel, Nehemiah, and Jeremiah—of God commanding his followers to leave bodies to decompose on the fields "like dung." In one conspicuous passage, Moses instructs the Israelites to pour the blood of slaughtered animals on the ground "because the blood is the life, and you must not eat the life with the meat." Return the blood to the ground,

Moses says, because "it may go well with you and your children after you." In Islam, the impurities of feces are considered cleansed by the process of decomposition, making compost an acceptable addition to the soil. Beginning in the eighth century CE, proponents of the Arab Agricultural Revolution advocated extensive manuring. The Muslim Arab botanist Ibn Bassâl, in the eleventh century, and the agriculturist Ibn al-Awwân, one hundred years later, carried out extensive testing of manuring and composting techniques. Today, fertilizing practices remain a part of Islamic cultures throughout the world—on organic farms in Israel, fish ponds in Java, and worm compost ventures in Sudan.

In 1596, Sir John Harington, a courtier to Elizabeth I and inventor of flush toilets, conveyed an allegory about an angel who prefers the smell of a gong farmer's dung cart to the perfumed aroma of a courtesan. At the time, Europe's population was growing, and cities were producing more raw sewage than ever before. Even though Western societies today are generally not associated with waste recycling and sustainable agricultural practice, such techniques were commonplace at that time. Soil samples taken around medieval archaeological sites in Russia have demonstrated heightened levels of phosphorus that steadily drop as the distance from the settlement increases—an indication that small-scale farmers were supplementing their soils with fertile goods. Concurrently, across Europe, archaeologists have found only minimal evidence of animal dung accumulation in cities, suggesting that manure was regularly moved into the countryside rather than left to rot in place.

In China, the application of natural phosphate fertilizers has long been a part of peasant practice. Ancient Chinese farmers employed dozens of varieties of organic fertilizer, including human waste, bird excrement, and livestock manure, within their carefully wrought systems of intercropping and rotational planting. They ground bones, composted human excrement, collected silt, and furrowed green plants back into the soil, maintaining a long-term balance between what went into the ground and what came out of it. Despite thousands of years of some of the most intense cultivation in the world, by the twentieth century, Chinese farmland was still fertile. Waste recycling also played an important role in the social progression of Japan. Boats bringing vegetables into seventeenth-century Osaka would depart filled with human waste, a two-way flow of nutrients that kept the land rich and the city fed. The provision of so-called night soil in Japan constituted an important part of tenancy: in some apartment complexes, tenants paid rent only in the form of their own excrement, which landlords then sold for profit. In other cases, rent might be raised if a household member moved away and deprived the landlord of his output. Throughout the seventeenth and eighteenth centuries, as fertilizer prices increased, the rules of collection were codified. Landlords controlled feces and tenants their own urine. At the time, the city of Edo may have been the largest in the world. Its growth into a major metropolis was premised on the efficient reallocation of human waste. In 1868, after the overthrow of the shogunate, the emperor moved the capital to Edo and changed the city's name: modern-day Tokyo was born.

Despite this long history of recycling, the relationship between life and chemical scarcity was not well understood by Europeans at the beginning of the nineteenth century. Before the makeup of chemical elements was established, the predominant theory explaining why plants grow as they do was that plants eat substances similar to themselves—that corn should be fertilized with decomposing corn, wheat with wheat, and so on. The German botanist Carl Sprengel was the first major scientific figure to suggest that all living beings require the same core materials in order to grow. Sprengel argued that the same few chemicals fed all plants—and that the growth of any plant, therefore, would be limited by the scarcest ingredient available. In the decades that followed, the great chemist Justus von Liebig developed Sprengel's work and popularized what became known as the law of the minimum.

Liebig's research was shaped by experiences of scarcity. Born in 1803, he turned twelve in 1815, the year a volcanic eruption in the Dutch East Indies caused temperatures to drop across the Northern Hemisphere. Summer failed to come in much of Europe, crops did not grow, and mass starvation ensued. It was the European continent's final experience of widespread hunger, and it inspired Liebig to science. As a child, he began experimenting with chemicals. He went on to study with the famed French chemist Joseph Louis Gay-Lussac, and he founded a scientific journal, *Annals of Chemistry,* at the age of twenty-eight. His interests were varied—he invented the food that would eventually be called Marmite—but his focus in those early years was on the interaction between life and its constituent elements. His research brought together chemistry and biology in ways that had not been tried before.

In Liebig's time, phosphate was an important but relatively unknown substance. Contained in organic materials, it had long been a part of farmers' fertilizing regimes, but its chemical identity had rarely been explored. In 1669, phosphorus had been isolated and named by the alchemist Hennig Brand, but no connection was made with fertilizer. In 1769—exactly one hundred years after Brand's discovery—the chemist Johan Gottlieb Gahn noted the phosphorus content of bones. Sprengel was born eighteen years later, and Liebig sixteen years after that.

Farmers faced a tough proposition. They were tasked with feeding a population that was growing and changing at a rapid rate. The rise of industrialism meant that farmers were expected to produce more food without occupying additional land—but, after centuries of cultivation, the soil of Europe's countryside was depleted. Gahn's discovery had boosted the international market for bones, the importation of which had begun in England at least as early as 1786. "Early representatives of the fertilizer industry scoured the killing fields of Leipzig, the Crimea, and of Waterloo too, to send back the bones of grenadiers and dragoons, hussars and cuirassiers, to be ground into powder and used to feed the new generation. In the military language of today, it's what one might call a collateral benefit of war," the geologist Jan Zalasiewicz wrote in 2011. By the 1820s, Britain was importing thirty thousand tons of bones every year.

The law of the minimum was formally postulated beginning in 1828, and it gave context to the important role of bones in raising yields. The main ingredient limiting the growth of crops, Liebig showed, was usually either phosphate or nitrate. Nitrate can be added to the ground by legumes, which fix the nitrogen in the air and return it to the soil: the widespread planting of clover in the later eighteenth century significantly alleviated nitrate deficiencies in many European soils. But phosphate is a mineral rather than a gas; it is not widely present in the air. As soils gained nitrogen, they continued to lose phosphorus. Liebig believed that by fixing this problem—by adding phosphate to the ground—struggling farmers could grow more food. Farmers responded to Liebig's research but faced another issue: although phosphate is present in bones, a ground-up bone remains insoluble. It is largely unusable by plants. The mass quantity of bones being imported was mostly going to waste. Phosphate can be applied to the land, but plants find it difficult to absorb. Getting around this quandary required one additional step.

The English aristocrat John Bennet Lawes was enraptured by Liebig's research. Wealthy and rebellious, he had, at the age of eight, inherited the Rothamsted estate, a thousand-acre property in Harpenden, outside London. He moved with his mother to Paris, where he helped build street barricades for the 1830 July Revolution, which overthrew the French king. Returning to England three years later, in 1833, Lawes enrolled in courses at Oxford. In his university classes, he was taught about the groundbreaking work of Sprengel and Liebig, about phosphorus scarcity and the law of the minimum. The rules of agriculture, it seemed, were about to shift. Lawes, intrigued, abandoned his studies a year later and returned to Rothamsted. In the six-hundred-year-old manor's historic brick mansion, Lawes began to carry out his own research, applying the lessons he had learned studying Liebig's work at Oxford. He dreamed of creating a fertilizer of epic proportions, one that would rewrite the rules of agriculture.

It started with a few potted plants, but Lawes's experiments quickly grew more complex. A visitor to Rothamsted would have found the venerable property overrun: the manor house held scores of plants, and the fields were divided to resemble growing conditions that ranged from undisturbed wilderness to immaculately fertilized soil. "I am afraid it sadly disturbed the peace of mind of my mother to see one of the best bedrooms in the house fitted up with stoves, retorts, and all the apparatus and reagents necessary for chemical research," Lawes would later recall. (He soon established a standalone chemistry laboratory in a nearby barn.) His goal was to make phosphate fertilizer work better. His ideas drew upon the works of others. In 1835, a German headmaster named Gotthold Escher had proposed moistening bone manure—essentially bonemeal—to make it decompose faster in soils, and in 1840 Liebig had reported to the British Association for the Advancement of Science that acid could, indeed, improve the solubility of bone manure. With these advances in mind, Lawes sought to invent a soluble phosphate fertilizer that could easily be applied on farms and absorbed by plants. In order to enact his vision, he dissolved phosphates in acid, applying the solution to test groups of plants.

The acid experiments worked. In 1842—the same year Henslow walked at Bawdsey—Lawes patented a solution of phosphate and sulfuric acid that was far more powerful than any other fertilizer used at that time. Lawes called it "artificial manure," but *superphosphate*—a name applied by Escher in 1835—would soon become the substance's accepted name. After the patent was filed, Lawes set up a factory to produce superphosphate for commercial use. Soon, many English entrepreneurs followed suit, starting with a man named Edward Packard, an experimenter who worked not far from Lawes in the southeast of England.

Like Lawes, Packard had been inspired by Liebig's research. He set up a superphosphate factory in the village of Snape, moving four years later to the town of Ipswich and then again to Bramford, a village near Ipswich through which a railroad ran. The year Packard opened his first factory was the same year—1843—that Henslow returned to Bawdsey to investigate the nodules he had found. The mining pits of Suffolk would go on to provide significant raw material for Packard's operation, which ground the rock into powder, added acid, and packaged the result. In 1851, Packard built a grand factory that streamlined the process of superphosphate production by carrying out all the steps—including making sulfuric acid—in-house. The Mills-Packard process of acid manufacture would become a global standard.

After seven years of operation in Bramford, Packard gained a competitor, a man named Joseph Fison, who built his own operation next door. Fison was the latest scion in a family of businessmen that started with his grandfather James, a baker. In Barningham, the family had benefited from the nationwide enclosure movement, obtaining as private property large portions of what had previously been common land. The money was concentrated in a family-run agricultural company that traded coal, made malt, and milled wheat. Following Lawes's discovery of superphosphate, the Fison family began importing bones from abroad in order to experiment with fertilizer production. By 1848—a year after the start of mining at Bawdsey—they were selling superphosphate made mainly from the Suffolk phosphate nodules. In 1858, Joseph Fison moved to Bramford to process phosphate. "I cannot understand why my Grandfather allowed the land next to his works to be bought by another fertilizer manufacturer," Packard's grandson, W. G. T. Packard, would later write in the company magazine, "but perhaps he did not fear competition." From 1858 onward, the competing firms were "separated only by a single wall."

As Packard's, Fisons, and a third company, Chapman Brothers, worked side by side in Bramford, they effectively industrialized the relationship between humans and phosphate. Their competition resulted in innovations that helped to move the industry forward. Joseph Fison specialized his firm's fertilizer offerings, selling unique dressings for wheat, barley, oats, turnips, and flax. He hired scientists to help create products that would make the biggest change. Customers could purchase fertilizers made of bones, of Peruvian guano, of coprolites. Meanwhile, Edward Packard was redesigning the process of the factory itself. He had figured out a way to fully enclose the factory's den, a white-hot chamber where acid and phosphate were combined, creating lumps of superphosphate but also noxious fumes that burned workers' lungs while the acid burned their skin. Through a series of tubes and fans that siphoned and filtered acidic air, Packard created the first fully enclosed den in England. He built a laboratory and eight acid chambers, a gasworks, and a crushing mill. By 1866, his works at Bramford were valued at £9,000. Three years later, he constructed thirty-two cottages, new acid chambers, workshops, and an engine and a mill on the property, and he purchased a barge to move fertilizers down the river. By 1889, he was mayor of Ipswich. The late University of Glasgow archivist Michael Moss described Packard's mayoralty in an unpublished book on the fertilizer manufacturers, writing that "as a tribute to the town he presented the museum with his private collection of crag fossils, including a magnificent tooth of Mastodon 'far surpassing anything of the kind in the British Museum or elsewhere.'"

The companies consolidated local political power, and they looked to expand beyond the region. After Suffolk, they turned their attention to neighboring Cambridgeshire. In the fens, those canal-crossed former swamps where Darwin played, a seam of phosphate had recently been discovered. In December 1847, the researcher T. Mainwaring Paine sent a sample of Greensand marl to a chemist, who identified the rock as high in phosphate of lime. In 1848, Henslow—having just successfully publicized the Suffolk nodules—turned again to *The Gardeners' Chronicle* to alert farmers to the discovery. By 1860, mining in Cambridgeshire was extensive. "If one walked from Cambridge along any of the roads into the country, within distances from the suburbs varying from two to three miles, the eye could not fail to see one, two, three or more pits in the adjacent fields," Henslow's son, George, later wrote. "The productive part of the band is about 8 to 12 inches thick and was reached by digging away the cover of Chalk Marl to a thickness which might amount to 20 feet," the researcher A. N. Gray wrote in *Phosphates and Superphosphate,* a 1944 history of the phosphate industry. "For the privilege of digging the coprolites about £140 per acre was paid and the average yield per acre was about 300 tons." The phosphate sold for £50 per ton. Packard invested £2,000 in a mine in Cambridge—equivalent to roughly $400,000 today—and many others followed suit. "There began a boom, a big expansion of the industry," Bernard O'Connor, the local historian, told me. The mining rested on back-breaking labor. Each man was expected to dig a ton of coprolites a day. At least one of the mines reached sixty feet in depth. "These pits progressed across the field. As the fossils were removed, their topsoil was then put back into the pit. And they would then work their way across. By the end of the working, the field was then one or two feet lower than it originally had been."

By 1862, according to George Henslow, phosphate nodules from around East Anglia—from Suffolk, Cambridgeshire, and Essex—were providing 72,000 tons of superphosphate per year, each mine dug painstakingly by men and boys. Injuries were common as heavy phosphates were carted across precariously placed wooden planks that rested on the walls of the pits. Feet slipped, walls caved, and people got hurt. But production grew, and the coprolites, sorted by young boys, were washed and taken to England's factories. By the 1870s, annual production peaked at 250,000 tons—but by then, the Bramford manufacturers were looking even farther afield.

During the 1870s, Packard's invested in phosphate mining in Spain and fertilizer manufacture in Ireland. J. E. Taylor, a geologist at the Ipswich Museum, helped Edward Packard Jr. find phosphate deposits in Germany and France, which were subsequently mined. Some of that phosphate was brought back to Bramford, while other amounts were turned into superphosphate at a new German factory opened by Packard's in 1873. A report at the time describing the Bramford works noted: "The coprolites are trimmed by men, from the huge heaps lying at the back of the mill, into a hopper which feeds a set of elevators. These take them up to a crusher, which, in its turn, delivers the crushed material to a pair of heavy stones, which further reduces them, and they are finally delivered by conveyors to the stones which finish the grinding. The ground or finished coprolites are caught in bags and taken to the store, where the mixers are fed." The mixers were the acids, and they tore apart their surroundings just as fervently as they broke the bonds of the phosphate that they treated: the factory, the report noted, was "lined with lead, which metal alone withstands the powerful acid, and even that is eaten away in the course of a few years." By the mid-1880s, the company's focus was on growing markets in Australia, New Zealand, South Africa, China, and the United States. Packard's was investing overseas as well, funding phosphate rock expeditions in Algeria, Canada, Florida, and West Africa. The Florida Phosphate Company, established in 1890, was funded by the Packard family. "Since Bedfordshire and Cambridgeshire coprolite mines were almost exhausted, the bulk of mineral phosphate was being imported from Algeria, the West Indies, France, the United States, and Belgium," Moss wrote. By the time Edward Packard Sr. died in 1899, the company's major superphosphate sales accounts included Russia, Martinique, and Japan.

The ledgers of the Lawes Chemical Manure Company—the first commercial superphosphate manufacturer, founded by John Bennet Lawes in 1843—reveal a decades-long shift in phosphate sourcing throughout this time. Pick a day—January 28, 1876. The board met and registered phosphate imports from the past month and a half. They registered two separate shipments of "whole Suffolk coprolites," totaling 450 tons, from Packard's. An additional shipment of 250 tons of Suffolk coprolites from Fisons. Three days later, an additional 70 tons of coprolites from Suffolk. Orders of bones, orders of guano—these are the standard of the early years. Throughout the 1870s, the company reported massive monthly shipments of bones—"bones from London," usually, at 50 tons, 340 tons, 500 tons. But the origins of those shipments began to change. On January 15, 1878, the company reported a shipment of "Foreign Bones," and, thirteen days later, a large shipment of South Carolina phosphate. The company's imports were becoming more geographically scattered: 110 tons of American bones on May 26, 1879, and 100 more the following February. Seventy tons of Egyptian bones, 30 tons of Moroccan bones. Bones from Bombay, from Calcutta, from Russia, from Odessa. The Lawes ledgers show the sources of rock as well: phosphate from Curaçao, from Belgium, from South Carolina, from France, from Canada, from Norway. Extremadura phosphate. Cambridge coprolites. Bedfordshire coprolites. Nauru, Ocean Island, Christmas Island, the Seychelles. Florida, Florida, Florida. Guano from Navassa, from Peru, from "fish." By the late 1870s, however, the Lawes Chemical Manure Company bore the name, but not the ownership, of John Bennet Lawes, who had sold it to investors in 1872. Lawes was back at Rothamsted performing research—and Justus von Liebig was dead.

. . .

In 1843, following Henslow's discovery of phosphate in the Red Crag, Liebig had been optimistic. He had written that "if a rich and cheap source of phosphate of lime were open to England, there can be no question that the importation of foreign corn might be altogether dispensed with after a short time." But throughout the 1840s and 1850s, as superphosphate production grew, Liebig felt frustrated. He too had discovered superphosphate, at the same time Lawes had, and in a similar way. And he too had opened a factory and sought to market the material. However, whereas Lawes's goal had been the advancement of agriculture, Liebig was motivated by the advancement of science itself. Lawes's fertilizers appealed to farmers, while Liebig's factory failed. Liebig was forced to watch the development of superphosphate from afar.

What he saw concerned him. Rapidly, Britain was scooping up the phosphate of the world. In 1841, Britain had begun importing guano from islands off the coast of Chile and Peru—the islands that had fed the Inca long before. "By the year 1844 the application of guano had become various and abundant," George Henslow wrote in a history of English fertilizers. Elsewhere, catacombs—cavernous burial sites situated underneath the great cities of Europe—collected phosphate that could not be ignored. "In these depositories of the dead, huge piles of human bones, from which all the animal matter had decayed before they were placed therein, are now reduced to phosphate of lime, the most valuable of all fertilizing materials," *Scientific American* wrote in 1857. "They emit no smell whatever." Later, when Britain conquered Egypt, a ship brought back 180,000 mummified cats that were auctioned in Liverpool and ground into fertilizer. In his final years of life, Liebig made clear his reservations about this system of phosphate importation. He accused the phosphate industry of engaging in a system of robbery—*raubbau*—of the soil. "Great Britain deprives all countries of the conditions of their fertility," Liebig wrote in 1862. "It has raked up the battle-fields in Leipsic, Waterloo, and the Crimea; it has consumed the bones of many generations accumulated in the catacombs of Sicily; and now annually destroys the food for a future generation of three millions and a half of people. Like a vampire it hangs on the breast of Europe, and even the world, sucking its life-blood without any real necessity or permanent gain for itself." Liebig worried that the world was wasting its phosphate. From rock deposits to factories to farms and people, from people to rivers and out into the sea, phosphate flowed. The changes were intensifying. What had once been a cycle of life and death would become a one-way flow from the ground to the sea, where the phosphate would be lost, seemingly, forever. The world had two choices, Liebig wrote: to deepen its reliance on phosphate rock or to find a more sustainable option. Liebig foresaw dangerous repercussions in the event that a more sustainable path was not chosen. "For their self-preservation, nations will be compelled to slaughter and destroy each other in cruel wars," he wrote, predicting that mayhem would break out, migrations would occur, and societies would collapse. "These are not vague and dark prophecies nor the dreams of a sick mind, for science does not prophesy, it calculates. It is not if, but when, that is uncertain."

Liz Harper's career as a paleobiologist began during childhood, at a fossil dig in Suffolk. Her father worked for Fisons. "There was a phosphate crisis, in what must have been the seventies, and Fisons couldn't get hold of phosphate. And my father was involved with, essentially, talking to the local geologists about whether it was possible to effectively reopen all those coprolite pits, and whether that was ever going to be a viable source of phosphate—rather than shipping it in from Africa. So he went around for one summer with the Ipswich geologists, digging holes, looking at these phosphate things here. And, of course, nothing ever came of it. But him doing that was really instrumental in making me a paleontologist."

By the time Doug Harper, Liz Harper's father, joined Fisons, the travels of his predecessors had opened England up to the world of phosphate rock. In 1929, the company absorbed Packard's, and in the years that followed, scattered employees sought to take superphosphate around the world. Far-flung correspondents, traveling for fertilizer sales or mineral prospecting, wrote home of lions in Rhodesia, wildflowers in Switzerland, and rainforests in Java. Walter G. T. Packard, who rose to become the vice chairman of Fisons, grew up wandering around his grandfather's factory in Bramford. In 1905, in order to investigate a guano deposit, Packard voyaged to the small Icelandic island of Eldey, a blocky monolith whose surface was so thickly covered with seabirds that it appeared from a distance to be capped with snow. Faced with the prospect "of obtaining skilled rock-climbers to help us up the sides of Eldey, I began to wonder whether scaling dangerous cliffs had been part of my bargain," Packard wrote, though he eventually completed a perilous ascent of the island's 250-foot sheer vertical face. He hired three rock climbers, who hammered iron spikes into the cliffs and shimmied up to various small ledges, and from each ledge a rope ladder was unfurled for Packard to climb. "When we seemed to have been climbing for hours and were only about half way up, it began to dawn on me that nothing short of a gold mine could make it worth while," Packard wrote of the experience. Soon after, the climbers tied a rope around the waist of their brave businessman and hoisted him the final sixty feet to the top. When the four men sampled the guano, they found it too thin to mine, and they left within an hour.

In 1961, when Fisons laid out its growth plans, there were so many areas of attention that the company's mission approximated global hegemony. The company wanted to focus on Commonwealth countries, on Latin America, on Europe, North America, Africa, Japan. The following year, fertilizer usage in Britain reached a level four times what it had been on the eve of World War II, two decades earlier. Farmers around the world had come to rely on Fisons for superphosphate, but in 1973 the industry faced a setback. A global oil embargo led to fears that suppliers of phosphate, like those of oil, might raise prices, and the company realized it needed more consistent sources of material. In an effort to solve this problem, the leadership looked to its past and wondered if the Red Crag, which had prompted the fertilizer industry to begin, might come back to save its future. Fisons wanted to know if phosphate in the Red Crag could feed the world again.

"I managed to get charged with the job of making sure we had no possible U.K. sources, which pointed to the old coprolite extraction," Doug Harper told me. "It was quite obvious from the outset that we were unlikely to be able to exploit U.K. phosphate sources because we felt that there would not be enough accessible and it would be unlikely to be suitable in any of our processes. But it was my remit to prove it." Harper turned to Bob Markham, a longtime curator at the Ipswich Museum and a man who possesses an unparalleled knowledge of East Anglian geology. The two men searched, Doug Harper told me, "rather as the people had in the nineteenth century"—by digging into pits along the edges of the London Clay, at the layer where lag would accumulate. Liz Harper would join some of those trips. In the years that followed, she attended Cambridge University as an undergraduate, then earned her PhD in mineralogy and biology at the Open University, studying oysters. She returned to Cambridge in 1992. She researched bygone organisms in Australia, New Zealand, Chile, France, and Hong Kong. In 2017 and 2018, she led the Sedgwick Museum, whose scores of cabinets and drawers contain thousands of priceless geological specimens, many of them excavated from nineteenth-century phosphate mines.

The elite universities of the United Kingdom were built, in part, by the flow of phosphate money. Cambridge was changed profoundly by the flood of phosphate that emerged nearby throughout the nineteenth century. Most of the major Cambridge colleges, and some of those at Oxford, participated in the extraction of phosphate. Between 1861 and 1873, St. John's College earned more than £12,000, an amount equal to $2 million today. In just five years, King's College—known then and now for its soaring chapel and Gothic architecture along the river Cam—earned £10,000. Christ's College, where Darwin studied, benefited more than most. Between 1867 and 1871, at the height of the mining, Christ's earned almost £6,000 from coprolite leases, an amount that constituted about a fifth of the college's total revenues during that time. Construction had boomed around the university. Just past Downing Street, paces from the airy cloisters of Emmanuel, Pembroke College rises in a looming mass of Casterton limestone imported from Rutland, to the northwest. These grand walls, which run along the road, create an air of enclosed regality in the middle of the town. The buildings were constructed between 1880 and 1883. During that time, Cambridge was changing. John Stevens Henslow had left the university largely because it failed to emphasize the natural sciences. His successors in the university's natural science faculties—supported by phosphate profits, inspired by students like Charles Darwin, and excited by recognition that Henslow's scientific study had helped create vast reserves of wealth—advocated for a transformation of the university's approach to the sciences. The results of that push, which included more funding to support students and an increased number of professorships, are most visible today in a sprawling complex of structures that populate the center of the city close to Market Hill. Just off Downing Street, near Pembroke, is the New Museums Site, constructed beginning in 1863. Across the road is the Downing Site, another relic of the mining era. Today the complex hosts museums and laboratories for subjects ranging from anthropology and archaeology to genetics, geography, and veterinary medicine. It is made up of looming, dark brick courtyards shaded by trees. The Sedgwick Museum is in one corner. It is a broad scientific monument that began construction in 1899, planned in the years that followed the death, in 1873, of Adam Sedgwick, a famed geologist and close colleague of Henslow's. The Sedgwick, and its city block of labyrinthine, intricate brick structures, was completed in 1904. It capped a multi-decade transformation of Cambridge—and its fossil and rock collections would preserve the record of phosphate that had traversed time. At the Sedgwick today, the study of phosphate continues. There are boxes and bags and drawers of phosphate collected over centuries.

Above the Sedgwick, in an upstairs laboratory, Harper and I looked at the insides of the rock Davies had pulled from the Bawdsey Cliff weeks before. We were peering through a microscope at a slice of rock one-thousandth of an inch thick, looking for evidence of its formation. What we found were worlds of preserved microscopic life. The slide, up close, looked like a savannah viewed from high altitude. There were amorphous patches of white, blobs of dark brown, tiny, granular flecks of varying shades of mud—a green, a gray, a tan—all set against a sea of particles that appeared almost orange. There was a sponge spicule—a needle-like piece of prehistoric sponge endoskeleton—a radiolarian that looked like a silhouette of a blowfish, and a series of sedimentary pieces. "It almost looks like those brown areas—the francolite, the phosphate, if that's what it is—are in blobs, which might be something about the diagenetic texture of it. It looks almost like a bunch of grapes," Harper said. The blobs revealed, in some sense, the formation of the phosphate: it had grown sequentially, in pieces, and then been bound together in this rock, where it compressed. It was a conglomerate of phosphate, a compacted collection of the essences of lives lived millions of years ago. It was a world preserved.

From that world—intensely normal but biodiverse, swelling with life but destined to become dry land—phosphate was created. Those lobsters, those ammonites, those clams and brachiopods and fish and turtles died and fossilized, or gave their phosphate to the life around them, or left it to disperse into the water. The early phosphate miners in Suffolk and Cambridgeshire capitalized on the processes of aquatic sedimentation, pulling fossils from the ground. From there an industry grew—first small, then large, then global, the industry of which Harper's father was a part. By the time Liz Harper got her start in paleontology, the phosphate industry was excavating the world. "My guess is that the motivation wasn't terribly malign. It's just something that grows. You've got a way of making money, and it just snowballs," she said. After the experiment when she was a child, when Bob Markham took Doug Harper to Bucklesham and the rock became superphosphate one more time, Fisons never made fertilizer with Suffolk coprolites ever again. The industry, long ago begun with the fossilized remains of prehistoric English life, had grown too big.

3

LIGHTBRINGER

By the middle of the seventeenth century, it had long been believed by various alchemists, astrologers, mythologists, and theologians that a substance called the philosopher's stone—which could heal ailments, dispense wisdom, and refine the human soul—had been created by God, given to Adam, and passed down perennially through multiple generations of men. The shaky history that supported this conviction imagined its own emergence to have come several millennia before it actually had, in 300 CE, when Zosimos of Panopolis first put the concept into writing. Its tenets, however, rested on a foundation of understandable belief. The historian Mircea Eliade wrote in his 1956 book *Forgerons et alchimistes* that stones have long been associated with fertility and love. The Phrygians worshipped Cybele as the Great Mother of the gods, but she also stood for mountains and mining. She was represented by a meteorite. The Greeks as a whole worshipped Eros, the god of love and desire, and they chose an unsculpted rock to represent Eros's physical form. At the center of the Great Mosque of Mecca is a black stone believed to have been cast by God to Adam, a scriptural meteorite that helped to build the Ka'bah, Islam's holiest site. Across world religions, earth is gendered female, the rocky planet fertile. The celestial origins of iron, which before smelting had come only from meteorites, contributed to the first known word for the metal, the Sumerian *an.bar,* meaning "sky-fire." When smelting was discovered and mining took on greater import, belief systems emerged around this seeming magical transition from rocky ground to useful tool. Mines were reproductive; ores were birthed like babies. Empty shafts regenerated. Sexuality was possessed by the ground. "There is the feeling of venturing into a domain which by rights does not belong to man—the subterranean world with its mysteries of mineral gestation which has been slowly taking its course in the bowels of the Earth-Mother," Eliade wrote of mining. Tools that smelted metal—the hammer, anvil, bellows—were venerated, the artisans revered as magicians. Eliade wrote, "It is the heavenly smith who fills the role of civilizing hero; he brings down grain from heaven and reveals agriculture to mankind." Jean Reynand, who had written two centuries earlier, was quoted by Eliade in the context of a natural alchemy. "As we make bread, so we will make metals," Reynand wrote, describing the work of harvests, the grinding of millstones, and the baking of dough. "Let us then co-operate with nature in its mineral as well as in its agricultural labours, and treasures will be opened up to all."

Fire doubles the truth of a substance, in successive stages. Light equates to sanctity. Minerals, materials, and substances can change and be changed through the acts of human hands. Life can be prolonged. Some number of years after God made the world, alchemists refined their methods. In makeshift laboratories set up in multiple countries, they worked to turn the earth into life. They wanted to find the Magnum Opus, the Great Work, the stone that God described to Adam, a mythological sky-fire. Refining and professionalizing their work, these astrologers became chemists. By the seventeenth century, they were largely serious in their research, balancing on the cusp of modern science. They were fully integrated into the priorities of society: somewhere along the way, the philosopher's stone had picked up wealth as another attribute—besides giving life, the stone was also believed to turn cheap substances into gold—and so, by the seventeenth century, chemical research was performed with the twin goals of making gold and achieving immortality. Hennig Brand chased both these goals in 1669 as he labored in his Hamburg laboratory, searching for clues to alchemical greatness. There was at the time a theory that urine, richly yellow, might contain an essence of gold. Brand worked the fluid hard, boiling and decanting it until the water was removed. What he found, however, was not gold but fire. The material he had isolated shone brightly. It glowed with white light. On contact with air, it leapt into flames. Brand named the substance *phosphorus,* a Greek word meaning "bringer of light." It was the first chemical element that had ever been identified.

The opportunities presented by Brand's discovery turned a search for gold into an investigation of a new product altogether: matches. Products purporting to create instant fire had been made for centuries. "At the slightest touch of fire they burst into flame," a Chinese writer had observed, seven hundred years before Brand, of a sulfur-based fire starter. Unlike sulfur, which requires heat in order to burn, phosphorus spontaneously combusts in air. Taking advantage of this attribute, European inventors after Brand began experimenting with phosphorus contraptions that could provide fire on demand—bottles or airtight jars in which the potent phosphorus was held safe until a hole was opened or a covering pierced, air let in, the phosphorus activated, and a fire set. These were cumbersome contraptions, however, and scientists looked for an easier solution. In 1826, an English chemist introduced friction lights to the market. A friction light was phosphorus on a stick, coated with paper that broke when struck, exposing the element to the air. In the early years of the nineteenth century, this new style of match gained popularity. The fires that lighted the English countryside—that burned Hitcham and inspired John Stevens Henslow to push for fertilizer from rock—were started with friction lights. The fires that prompted superphosphate had been lit by phosphorus itself.

In the ashes of those fires, a shift in agriculture reordered the world. The shining light of phosphorus, the promise of a future in the patterns of the ground, activated in humanity a feeling of idealism. The paleontologist William Buckland, who served as Liebig's guide on a trip to scavenge coprolites in Devon, waxed romantic about the new discoveries: "In all these various formations our Coprolites form records of warfare, waged by successive generations of inhabitants of our planet on one another: the imperishable phosphate of lime, derived from their digested skeletons, has become embalmed in the substance and foundations of the everlasting hills; and the general law of Nature which bids all to eat and be eaten in their turn, is shown to have been co-extensive with animal existence upon our globe." The cycles of life, Buckland argued, were embalmed in the record of the rock. "It follows," he wrote, "that wherever decomposition of foecal dejections, or of dead animals, or of sea weed, was going on over the entire bottom of all the ancient seas and great ocean beds—these cumulative additions in every day and every hour of antediluvian time, became the grand receiver-general and cloaca maxima of the terraqueous globe—an universal laboratory and conservative store house of universally dispersed manurance for the future cornfields and pastures of the earth, when the ancient sea-bottoms should be raised up to become dry lands."

Phosphate fertilizers were part of a new agricultural revolution that sought to treat the land better while improving the efficiency of farming. Crop rotations prevented the nutrient-sucking effects of repeated cropping, while cover crops, known to protect soil health, were introduced. Legumes, such as clover, enriched the soil by fixing nitrogen from the air, moving it to solid form. In England, the cultivation of legumes increased after the mid-eighteenth century, relieving the soil of one limiting factor but making phosphorus an even greater bottleneck for farmers. Superphosphate relaxed that restriction, and the countries that applied the most fertilizers generally achieved the highest levels of productivity. The application of superphosphate raised yields at a critical moment for industrial development and population growth. "If the commercial manufacturing process for phosphatic fertiliser had not been discovered in the 1840s, life in the UK and in other industrialised countries would have become unsustainable," two researchers wrote in an article commissioned by the United Kingdom's Government Office for Science in 2011.

In agricultural science, the relationship between inputs and outputs can be difficult to parse. The processes of growing plants are complicated and interconnected—but following the introduction of superphosphate, certain changes were difficult to ignore. The yields of wheat in England, which hovered around one ton per acre in the year 1800, increased sixfold after that. According to a 1935 analysis by Stanford University's Food Research Institute, the century that elapsed between 1750 and 1850 saw an increase in wheat yields that amounted to more than that of any other one-hundred-year period in what was then a seven-century history of wheat yield statistics in Britain. The measurement of phosphate fertilizer added to farmland is often given in terms of kilograms per hectare. One kilogram per hectare of land would be roughly equivalent to the weight of a marble spread across every five hundred square feet of land. By 1870, fewer than thirty years after superphosphate was first patented, the application of artificial fertilizers in Britain had reached a level of five kilograms per hectare of farmed land. By 1880, the number had increased to seven. The only country that applied fertilizers at a higher rate was Belgium, where farmers were applying nine kilograms per hectare every year. In 1870, Belgium's farmers had the highest yields in all of Europe. They were followed, in descending order, by the Netherlands, Britain, and Denmark. Where fertilizers had been applied, more food had been grown, and the nature of farming had changed.

Phosphate was consumed, crops were raised, new kinds of farming were practiced, and massive amounts of food were produced throughout Europe. Populations grew. Fewer people needed to farm. Meanwhile, new lands were opened up abroad, lands that had not yet been stripped bare of phosphate. In the prairies of North America, the bison had been killed and their bones, in many cases, exported to Europe to make phosphate fertilizers. In England, there was a new reliance on imports, particularly of wheat and butter, from continental Europe but also from New Zealand and the United States. It was a profound process of extraction of phosphate from the ground, a vast global movement of nutrients from distant land to domestic bodies.

These changes began in England, and England was first to see their next effect: as the century continued, British cities surged. Phosphate fueled the Industrial Revolution. Between 1801 and 1901, the population of London increased nearly sevenfold; its physical footprint grew from just a few square miles at the beginning of the century to nearly seven hundred by the end of it. By that time, thirty-six cities throughout Britain had populations that surpassed one hundred thousand, a feat that had been achieved by only one—London—a century before.

The lore of London is closely tied to this period of its growth. The Great Exhibition, the world's first world's fair, took place in 1851. The Metropolitan Railway—the first line of what became the London Underground—opened in 1863. Charles Dickens's life spanned 1812 to 1870; his career followed the rise of superphosphate in England, and his work chronicled the effects of city life on the urban poor. One of the central problems of the time was the effect of crowding on sanitation. At the beginning of the nineteenth century, human waste in London was generally collected in cesspools, pits that were emptied at night, with excreta taken to the countryside as fertilizer. This system worked until the population of London exploded and a dearth of regulation resulted in widespread leaking, overflows, and structural failures. The introduction of flush toilets early in the nineteenth century doomed the cesspools to oblivion: the pits could handle waste, but they were unable to cope with the large amounts of water that increasingly were used to carry waste out of houses. Sewage accumulated in the city, creating waves of cholera that caused thousands of people to die. The Thames, polluted, stank of death.

In 1828, the painter John Martin proposed a centralized sewer system that would turn solid waste into marketable fertilizers. Six years later, the civil servant Edwin Chadwick, in his more practical push, argued that liquid waste, pumped into the countryside and sold to farmers, could help fund the system overall. The government tasked Chadwick with further designing plans, and as he guided London's engineers and public health officials toward a new sewage system, the idea of fertilizer output remained an important part of the plan. Joseph Bazalgette, the engineer who designed London's Hammersmith Bridge, was hired to design the sewage system. In

1849, Bazalgette proposed a system of public toilets whose operation would be supported by the sale of fertilizers made from urine. For disposing of solid waste, the creation of fertilizer cakes was proposed. But those ideas were derided as uneconomical, and the idea of recycling sewage fell out of favor. Over nearly two decades, as Bazalgette built a sewage system in London, the details of his original proposal gradually fell away. By the time the system opened, in 1865, the connection between waste and farming had been lost entirely. Sludge from the city was taken to the coast, loaded onto a rotating cast of ships, and dropped—at a rate of two thousand shiploads per year—into the North Sea.

Other cities quickly followed suit. The year before London's system opened, the government of Paris began a forty-five-year project to construct a sewage system of its own. A similar twenty-five-year project began in Naples in 1889. There were projects in Lyon and Marseille, in Brussels, in Florence, in La Spezia. Some of these entailed replacing not just a sewage system, but an entire city. The late nineteenth-century rebuilding of Paris, a project that opened up wide avenues and introduced fresh grandeur, was also, in part, a public health measure: the boulevards were built atop new pipes that centralized disposal of the city's waste. Flush toilets and the riverine disposal of human excrements became a hygienic expectation throughout the world. The aims were noble, but their implementation was out of line with the processes of nature. The phosphorus cycle functions only if nutrients taken up by organisms are returned to the ground for the benefit of the organisms that follow. ("The blood is the life," Moses had said.) Increasingly, by the end of the nineteenth century, human waste no longer followed this path. Phosphate, taken from the land, was charting a rapid course into the sea.

While some chemists were experimenting with the creation of fire, others were beginning to understand the meaning of it. In England, Joseph Priestley investigated the relationships between components of the air. Priestley, in the tradition of the time, was a researcher of varied interests. He was a serious practitioner of chemistry, particularly gas and liquid, and is noted today as the inventor of carbonated water, the precursor to all modern fizzy drinks. He was also a Dissenter, a religious radical opposed to the primacy of the Church of England, and his belief system was wide-ranging. Debate at the time was centered on the relationship between matter and life. Philosophers such as John Locke, John Stuart Mill, and David Hume were asking questions about meaning and morality, about matter and the soul. *What is the meaning of meaning?* these philosophers asked. *What is the basis of thought?* Priestley was an avid participant in these discussions. Within them, he carved out a space of scientific practicality. He accepted the tenets of associationism, which held that physical sensations define the contours of human thought. A theologian as well as a chemist, Priestley believed that consciousness ended with death. The soul, informed by its surroundings, was a temporary being that would one day disappear.

It was common knowledge among chemists at the time that air interacted with life in complicated ways. Animals breathed air, changing its fundamental nature. Experiments showed that plants grown in closed systems with water and air increased their mass by more than the mass of the water they absorbed—suggesting that some material came from the gases around them. Fire, too, was a part of this equation: without air, fire could not exist. Priestley believed that all of these observations were connected. In 1771, he began putting mint plants into bell jars in which candles had burned the air, and he found that the plants restored the air to its former state, allowing a fresh candle to burn. He had already found that a mint plant, growing beside a flame, would keep the flame burning, and that a mint plant, in a jar that also held a mouse, would keep the mouse alive. Throughout the year that followed, he tested combinations of these variables, burning different types of candles, metals, and coal, and observing the abilities of flames to burn, mice to live, and plants to rejuvenate the air. By 1772 he was able to write, drawing on these experiments, "that the injury which is continually done to the atmosphere by the respiration of such a number of animals, and the putrefaction of such masses of both vegetable and animal matter, is, in part at least, repaired by the vegetable creation."

In the autumn of 1774, Priestley left Britain and traveled to continental Europe, where he dined with the French chemist Antoine Lavoisier. A famed experimenter who is now considered to have invented the modern discipline of chemistry, Lavoisier was a powerful and well-regarded figure of his time. He was known for conducting intense experiments, including ones that put his body at risk, as well as for championing the metric system and developing a new kind of chemical nomenclature that is still in use today. Lavoisier performed his research with his wife, Marie-Anne Paulze, a figure whose life story exposes the complicated social dynamics of the time. Paulze had been married to Lavoisier as a child, a union of convenience that was unsettling but also strategic: the two had been wedded in part to prevent her marriage to an even older man. Paulze and Lavoisier worked closely together, but he has been the one to receive the credit for their discoveries. A talented chemist and translator, Paulze's status as a woman meant that she was mainly regarded as a valuable laboratory assistant.

Throughout the 1770s, Paulze and Lavoisier had been conducting their own experiments with fire. Around the same time Priestley was putting mice in bell jars, the French duo had been lighting fires and trying to understand why they burned. They had noticed, in 1772, that phosphorus—elemental and instantly flammable—increased in mass when it burned. Later they found that sulfur reacted similarly. They proposed that the added mass originated in the surrounding air, evidence of a change in matter that paralleled the gas exchanges Priestley had found between the mint plant and the candle. We now know that phosphorus, burning, was taking oxygen from the air.

Paulze had translated Priestley's work from English into French, so Lavoisier was familiar with it. When he met with Priestley in 1774, they discussed Priestley's experiments. They talked about the mouse and the mint and the flame. In those experiments, Priestley had discovered an air he felt was purer, and he described its properties to Lavoisier. Candles burned more strongly in this substance, and mice lived twice as long. Priestley called it "dephlogisticated air," but Lavoisier, in years that followed, would name the element *oxygen.*

This element, the chemists realized, was present in the air but also crucial to the bodies of living organisms.

Talking shop in Paris, Priestley and Lavoisier shared a common outlook on the world. In an age of revolution, both men believed in change. Priestley advocated against the state-sponsored church. Lavoisier supported reform and sympathized with those who opposed monarchic rule. Both men supported the American Revolution: the British Priestley opposed the use of force against the colonists, and Lavoisier, an official of the French government, would go on to engineer the provision of gunpowder to the American revolutionaries. These feelings of revolution extended into science. Since the time of Anaxagoras, 2,100 years before, scientists had recognized the immutability of matter. Throughout the 1770s, Lavoisier conducted experiments that probed this recognition in the context of life and fire. In the 1780s, he worked with the polymath Pierre-Simon Laplace to develop an ice calorimeter, a chronicler of energy that revealed the startling similarity between a breathing mouse and a piece of burning charcoal: they both produced the same gas. In 1785, he formulated the law of conservation of mass, stating that matter could not be created or destroyed but only changed in style from one substance into another. In 1789, when he published *Traité élémentaire,* the theory came together in full. "We may lay it down as an incontestible axiom, that, in all the operations of art and nature, nothing is created; an equal quantity of matter exists both before and after the experiment; the quality and quantity of the elements remain precisely the same; and nothing takes place beyond changes and modifications in the combination of these elements." It was a law of equality. In an experiment or on earth, the beginning and end would be the same. Nothing would be lost, and nothing gained.

The cycles of matter were coming into focus. The experiments conducted by Priestley, Paulze, and Lavoisier showed that the consumption of oxygen by the candle and the consumption of oxygen by the mouse were, essentially, the same process—they showed that all of life is the product of a burning flame. The modern notion of the carbon cycle rests upon these observations. Living matter is continuously being produced and then destroyed. The elements that make up our bodies—carbon, hydrogen, nitrogen, oxygen, phosphorus, and others—are constantly changing, are being transformed into new states of matter, are moving between life-forms. Plants photosynthesize, turning carbon dioxide and water into oxygen and glucose. Respiration burns oxygen and turns it into carbon dioxide, providing energy for bodies of life. The candle, in the bell jar, turns oxygen back into carbon dioxide, feeding the plant. If the plant is absent, the mouse dies.

The oldest known phosphate rock dates from the Paleoproterozoic, an era of geological history that began 2.5 billion years ago. The defining feature of the Proterozoic, the eon of which the Paleoproterozoic is the first part, was something known today as the Great Oxygenation Event, so called because it was the time when free oxygen gas first began to populate the atmosphere. For 2 billion years, earth's atmosphere had been rich in nitrogen and carbon dioxide but low in pure oxygen, which was, for the most part, held captive in the crystalline structures of igneous rock (and in water). The rise of early oxygen-producing photosynthetic organisms began to disrupt this balance, ultimately resulting in the breaking up of water molecules and the releasing of their oxygen as gas. (The rise, 2.5 billion years ago, of these oxygen-producing photosynthetic organisms is not to be confused with the advent of other, anoxygenic organisms, about a billion years earlier, that practiced a kind of photosynthesis that does not create oxygen.) Slowly, earth's atmosphere accumulated the oxygen released by this new kind of photosynthesis, attaining higher oxygen levels that have continued to the present day. Since then, our atmosphere has been oxidizing: it pulls electrons from compounds, it rips apart rocks—it rusts and rots and inflames the other elements of the earth.

Increased weathering of rocks released sodium, sulfur, potassium, and phosphorus. Moreover, and perhaps more importantly, the large-scale transformation of carbon dioxide into glucose, via the photosynthetic splitting of water, fundamentally enhanced the availability of energetic electrons to the rest of the biosphere. As a result, life flourished during this time. Single-celled organisms began to form more complicated structures. Stromatolites, massive, layered domes of mineral-encrusted algae, proliferated. Other forms of photosynthetic life, ancient and diffuse, dot the fossil record of this time, growing in size and complexity. Besides oxygen, fossils, and the odd Australian stromatolite, the main legacies of the Proterozoic are deposits of minerals involved with living organisms. Phosphate deposits from this era have been found in Brazil and Northwest Africa and across central and Southeast Asia, with some others spread throughout Europe, North America, and Australia. This was the earliest preserved episode of phosphate rock formation in earth's history: for the first time, phosphate was abundantly cycled in the living world, and life consumed it and collected it in a significant way.

As the Precambrian gave way to the Phanerozoic, the oceans rose and circulated at a faster rate, bringing phosphate from deeper parts to the surface. Plant and animal life surged. Fed by melting glaciers and the breakup of the supercontinent Rodinia, rising water submerged long-dry land all over the world, creating shallow seas where phosphate could collect. Phosphate rock began to form en masse—deposits in central and Southeast Asia and in Australia, with smaller deposits left in Europe and North America. Phosphate deposition peaked around 540 million years ago, in the middle of the Cambrian period, as life flourished in the oceans. Much of the known fossil record begins at this time, and it displays a startling rate of growth. The scale of change is difficult to overstate. Throughout the earth's first 4 billion years, living organisms had been small and limited. Phosphate fed the Cambrian explosion—just as it later did the Industrial Revolution in London—and life evolved new traits. The world was warmer, wetter, and more diverse. The early paleontologists, including Buckland and Darwin, had noticed this change, with some postulating that life itself had started in the Cambrian. Life had not started, but the animal kingdom had; organisms had begun to develop skeletons, insects had come to the fore, and Adam Sedgwick, Henslow's colleague at Cambridge—and the namesake of the Sedgwick Museum—gave the period its name.

There had been great fluxes of nutrients within the oceans. Spurred by this fertilization, algae flourished, pumping oxygen into the air. The biomass of the earth increased. Some of that life was preserved in stone: by the time cooling temperatures brought about the end of the Cambrian, billions of tons of phosphate had been secreted in basins all around the world. In 1875, the geologist Harry Hess found bits of phosphate preserved from the Cambrian near Cader Idris, North Wales, and in Pembrokeshire; in 1893, the New Brunswicker William Diller Matthew, a future longtime curator of the American Museum of Natural History, discovered two deposits near his hometown of Saint John. J. G. Andersson, in 1896, found two Cambrian-era phosphate deposits in Sweden, one near Örebro, in the south, and the other on the island of Gotland. These were tiny bits of phosphate, clues in the search for larger deposits, but not themselves significant sources of phosphate rock—they were not mined. In the twentieth century, however, the clues would accumulate, the discoveries multiply. The life of the Cambrian had been preserved in all these little pockets, the seeds of modernity buried below the ground.

The Ordovician followed the Cambrian and was typified by an expansion of life—the first land plants, the growth of vertebrates—and, in the end, a mass extinction. In the northeast of Estonia, along a fat line that extends just south of the Gulf of Finland, geologists have found substantial phosphate rock. In central Tennessee, a north-south bed of phosphate runs from Alabama nearly to the Kentucky border. That phosphate infiltrated and replaced the limestone bedrock, forming a deposit that was mined from the end of the nineteenth century to the end of the twentieth. The Ordovician also left behind a smattering of smaller deposits in Canada, Bolivia, and Australia, perturbations of geology that were followed, for the next 250 million years, by long, phosphogenic silence. The Silurian, Devonian, and Carboniferous periods passed. Coral reefs grew, fish multiplied, swamps transformed to coal—and the Permian began.

At the outset of the Permian period, nearly 300 million years ago, the continents had collided to form Pangaea, sea levels had dropped, and the temperature of the earth had fallen far. During the Permian, the air warmed and seas rose. But the oceans bled oxygen, and they subsequently deteriorated into an acidic mixture that proved toxic to much of life. Something of an apocalypse occurred at the end of the Permian, and nine out of ten species disappeared. The death of life accumulated. The remnant of the grandeur of the Permian would be a single massive phosphate deposit in the Tethys Sea. In the cold water of that sea had lived fish, snails, bivalve mollusks, eel-like conodonts, and algae, all of which are preserved in the phosphate deposits that exist today. Those organisms rested on the bottom of the sea, where they were decomposed by swarms of bacteria that released the phosphate from the organisms' bodies and allowed it to coalesce into rocks that encapsulated, or sometimes replaced, the remains of the organisms that had not decomposed: the fossils that remain today. The deposit has been named Phosphoria, and the fossils that are found inside it range in age from 284 to 252 million years, an extent of time that ends with mass extinction. The reason for that extinction may, in part, have been a dearth of phosphate in the oceans. In the western United States—in Idaho and Utah, Wyoming and Montana—the Phosphoria Formation preserves the phosphate that the oceans may have missed. Much of that rock is inaccessible, buried thousands of feet beneath the ground or elevated into mountains. Only some small sections of Phosphoria, hidden near the surfaces of certain dipping slopes, can be mined.

The Triassic followed the Permian, and the Jurassic followed that. Pangaea split apart during the Jurassic period; the world warmed. Dinosaurs walked on land, and giant sea creatures swam the waters. Phosphate was deposited in modern-day Russia, in Pakistan, in Australia, and in other spots scattered throughout Europe and the Americas. Dinosaurs reached their apex, and the Cretaceous period dawned. During the Cretaceous, dinosaurs roamed. Pangaea continued to break apart, and the separated continents revealed more edges, more spaces for phosphate to accrue—across Europe and Australia, in pockets throughout the Americas, in Kazakhstan, in Pakistan, in Texas. The largest of these deposits formed along the coast of North Africa, where today there are Morocco, Western Sahara, and Algeria. Of these deposits, several things are true. They are larger than any but Phosphoria. They are formed of ancient coasts and inland seas. They are younger. They sit closer to the surface. They are the largest accessible phosphate rock deposits in the world. And they were formed just as an extinction, the Cretaceous-Paleogene, wiped out much of life on earth.

During the Cretaceous, some phosphate accumulated around Cambridge, in the fens. After that, the Paleocene began, and then the Eocene, which started 56 million years ago. It was the beginning, in some ways, of the modern era: the earth began to look the way it does today. The London Clay Sea was large, and those fateful nodules were deposited at Bawdsey. Globally, however, the Eocene did not play host to a major phosphogenic episode. It ended, and the world changed yet again. The ice melted, the oceans rose, and 14 million years ago, along the watery coastlines of the world, life flourished. In California, Venezuela, Japan, the Philippines, and, most significantly, in Florida, a final bed of phosphate rock was laid.

As they worked to change the rules of chemistry, Lavoisier and Priestley adopted the language of the upheaval that characterized their time. As cities grew and democracies toppled monarchic rule, the old alchemical order was replaced by a modern chemical system organized through conceptions of separation and combination, of the constant breaking down of matter into tiny parts to be assembled anew. It was a revolution in chemistry, and it advanced side by side with the revolutions that challenged the powers of Europe. But these different kinds of change did not, in the end, reflect one another. "Revolution as Lavoisier saw it had the stately inevitability of planets revolving around the sun. The streets of Paris running with blood did not enter into his vision. The replacement of one order by another was revolution enough for him. He did not foresee an order succeeded by anarchy," Madison Smartt Bell wrote in a 2005 biography. In Lavoisier's mind, Bell wrote, a revolution should be well considered—careful, determined, and decidedly not violent.

In 1789, as the French Revolution grew, Lavoisier was caught atop a powder keg. In all his experiments with fire and explosives, he had set himself up in the Arsenal, the country's gunpowder store, which he oversaw as the commissioner of the Gunpowder Administration. His home and laboratory were located in a preening symbol of French monarchy, and his positions within the French government, including as a tax collector, made him a prime target of the citizens who were agitating for change. In June 1789, the king allowed the National Assembly to form. When an uprising still seemed near, he ordered gunpowder be consolidated in the Bastille. Lavoisier, the chemist of combustion, was asked to carry out this order. Two days later, the Bastille fell to revolutionaries. There was rioting in the streets. The monarchy crumbled and was replaced, largely, by wealthy reformers, elites who were more independent of the nobility and the clergy. But the changes were not enough, the movement became more radical, and the revolution's leadership turned against its monied men. They declared a French Republic in 1792, and extremists consumed the government. The king was executed. In 1793, the Reign of Terror began. Lavoisier, who had profited wildly from a government tax-farming scam, was arrested for corruption and tried before a revolutionary judge. An accidental character in an age of revolution, he was executed by guillotine in 1794.

In 1794, too, Priestley's country turned against him. He had, three years earlier, been driven out of Birmingham by a mob that burned his laboratory to the ground. Hounded wildly for his avowed heretical beliefs, Priestley moved to America just as Lavoisier was killed. Nine years later, Justus von Liebig was born. In 1842, when phosphate was found in England, Liebig would understand the gravity of the discovery in the context of the histories of chemistry. "In the remains of an extinct *animal* world, England is to find the means of increasing her wealth in agricultural produce," he would write. "May this expectation be realized, and may her excellent population be thus redeemed from poverty and misery!"

Perhaps the greatest consequence of Henslow's discovery in Suffolk had been the global search itself. Following Suffolk in 1847, there was mining in Norway in 1851, in France and the West Indies in 1856, and in the United States and Germany twelve years later. In the 1870s, mines were opened in Canada, Curaçao, Belgium, and Spain. The 1880s saw the beginning of mining in Aruba and Russia, as well as the opening of an Algerian mine, the first one in North Africa. There would soon be mines in Redonda, French Guiana, Sweden, and Tunisia. Christmas Island, three hundred miles from Jakarta, was the final new phosphate mining operation of the nineteenth century.

The beginning of the twentieth century saw the opening of a slate of phosphate mines across the Pacific—on the neighboring islands of Banaba and Nauru in the Pacific, in New Zealand, in South Australia, in Japan, on Angaur, on Makatea, on New Caledonia. Then there were mines opened in Madagascar, India, Portugal, China, Morocco, the Philippines, Estonia, Ireland, Poland, Romania—and the list continued. Many of these were smaller deposits, individual mines crowded onto smaller seams of phosphate that were quickly consumed following their discoveries. In 1913, a year before the outbreak of World War I, global phosphate rock production peaked at more than 5.5 million tons. Production picked up after the war and reached 12 million tons before the Great Depression again reduced production. This global network of mines on islands and along rivers, by coastlines and in quarried-out pits inland, brought past lives together to create future prosperity—a global trade that resurrected history.

Fisons pioneered this trade, then dominated it, then disappeared. In the middle of the twentieth century, the company diversified its holdings in a variety of chemical industries, and in 1982 it sold its fertilizer business to the Norwegian chemical company Norsk Hydro, which later broke off its fertilizer holdings under the name Yara International, which is now the fourth-largest fertilizer company in the world. In 1995, the remainder of Fisons was acquired by the French pharmaceutical manufacturer Rhône-Poulenc. A year later it was sold again, to Medeva, a pharmaceutical company that soon ceased to exist. Throughout these changes, the factory on the edge of Bramford, which had been started by Packard and then swallowed by Fison, continued to produce fertilizers until 2003. The factory had spanned nearly the entire phosphate era. In the years after production stopped, it was acquired by a developer, and it fell into disrepair. There were holes in the roof, strips of metal flying off during storms. Six fires had been reported at the site. Kelvin Dakin, who had grown up in Bramford and still lived there, launched a petition drive to save it, and in January 2019, the Mid Suffolk District

Council, which exerts planning authority over Bramford, obtained a court order forcing the property's owner, Paper Mill Lane Properties, to take care of the site. If the warehouse roof was not fixed by the end of July, the order stated, the council would take enforcement action. Months passed, and the roof remained unfixed. Then, on the morning of the sixth of May, two and a half months before the enforcement was set to take place, Dakin received a call from the local television station. The warehouse was on fire.

The fire had started in the night—details are blurry—but people nearby heard explosions before sunrise. The fire department was called at 4:59 a.m., and over a period of an hour, sixty-seven firefighters arrived on the scene in eighteen different vehicles, the first arriving seven minutes after the call. By dawn the site was an inferno, intense bowls of brightness, of yellow and red, giving way to black smoke that entered the sky in plumes. The flames could be seen across Bramford. The explosions were heard as far away as central Ipswich, two and a half miles to the east. Smoke was visible throughout the region. The entire building was covered in flames, and the rooftop—which had long since peeled—was now gone entirely. Pieces of metal flew onto neighboring properties, causing damage that residents would later report cost the equivalent of more than a quarter million dollars to repair. "There were embers bigger than the size of your hand falling," Samantha Pemberton, who lived near the warehouse, told the BBC. "It was like walking onto the set of a disaster film."

Two years after the fire, Dakin and I visited the site. The factory looked as it had just after the fire—charred, emaciated, bleak. We climbed through overgrowth along a path that perched above the river, below a fence. At the spot where docks had sat, we climbed over, and then we went up, level with the train tracks, and emerged into a clearing. We were behind the old grinders—where the coprolites had once been turned to dust, then moved across the way and turned to superphosphate, then shipped away. "The reason Packard, Chapman, and Fison came here was because of the rail link. It didn't matter that Packard wasn't in Ipswich right on the docks—he could easily transport raw materials out by rail. But in the 1860s, the railway companies were thought to be a bit unreliable, and charging what they thought was too much money. So it was then, in 1868, that Packard launched his first steam barge on the Orwell, trialed it down to Ipswich, and thereby started his own barge traffic. It must have worked out, because they ran the barges for at least sixty years." The superphosphate that was made here, first by Packard's and then by Fisons, redistributed phosphate around the world. This factory accumulated rock from Suffolk, from Cambridgeshire, from Morocco and Algeria and Florida and South Carolina and Nauru and Banaba, and it sent that material throughout Britain, Europe, and the world. It changed agriculture, and it changed society. And then, in the brightness of a night illuminated, it disappeared—destroyed, according to the fire department's official report, by "heat source and combustibles brought together deliberately." Tanks of liquefied petroleum gas were found by firefighters on the scene. No suspects were identified, but Paper Mill Lane Properties, facing enforcement action over the site's preservation, was seemingly the only beneficiary of the arson. A representative of Paper Mill did not respond to multiple requests for comment for this book. In 2023, the property was put up for sale.

We made our way downhill, through the thorns of wild roses, to the river. The water below us rippled in a gentle wind. Lily pads floated, birds sang, and elderberries drooped overhead. We walked back to the gate, and a kestrel flew above us. "Although the fertilizer industry in Suffolk wasn't born here, it certainly grew up here," Dakin said. This factory nurtured a global industry. ("Difficult to imagine now what it must have been like.") Cracked concrete spread out before us, brick ruins, graffiti: the grim equality of destruction by fire. The air was warm. Signs attached to fences urged outsiders away. We peered inside at the skeleton of the building, around which flowers grew.

II. GROWTH

THE COUNTIES *of* THE BONE VALLEY FORMATION *of* FLORIDA

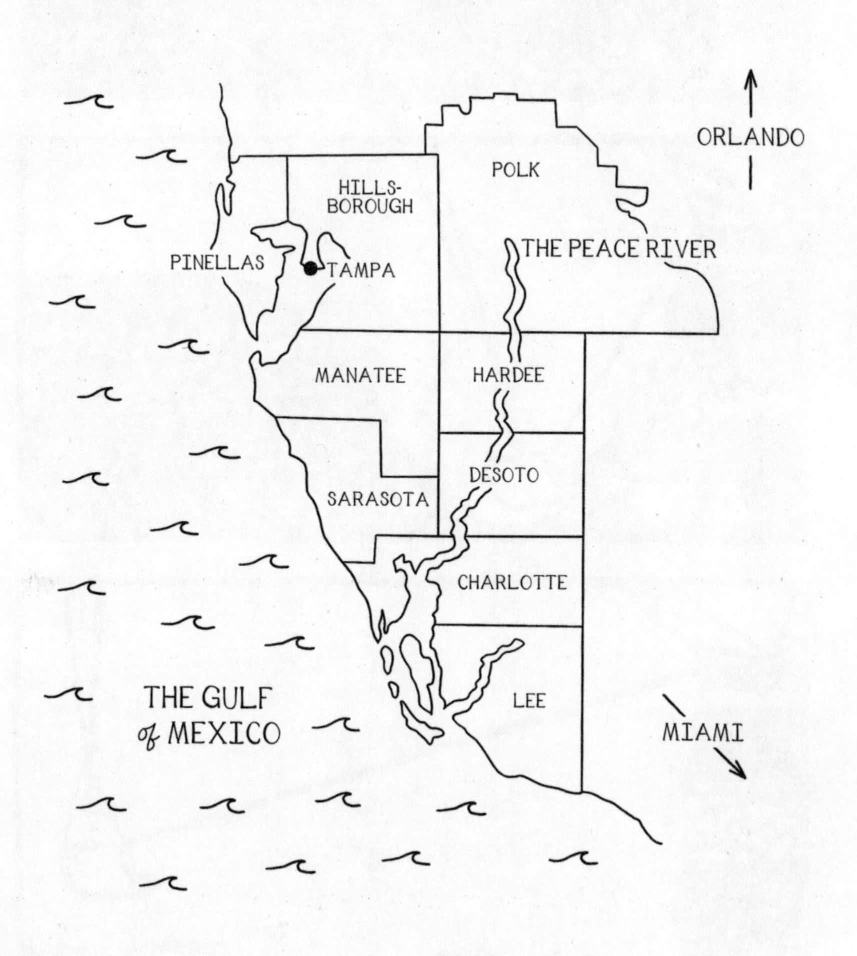

4

STONES FROM PAST TIMES

Thom Downs is not a character. He is soft-spoken, and he avoids making sweeping statements. He understates his experience, but he speaks with quiet authority, and he is not given to moving sharply—so when he jumped back, pulled his hand from the murky water that surrounded us, dumped a sieve full of gravel, and looked alarmed, I jumped, too.

"Was that a fish?" I asked him.

He did not know.

It was morning, early spring, and the air was hot already: this was Florida. We were walking the Peace River, a loopy, swampy thread that stretches from the lakes of Polk County—due east of Tampa—south to Punta Gorda, spitting out its contents into Charlotte Harbor and, eventually, the Gulf of Mexico. We were in rural, relatively unpopulated Hardee County. A landlocked rectangle crisscrossed by creeks that eventually, for the most part, make their way to the Peace, Hardee is a place that traffics mainly in citrus and cows. Wide at its mouth, the Peace narrows substantially farther up. Where we stood, it was twenty feet across and muddy, variously fast-moving and slow. It reached our stomachs. We walked down the center of the river, veering to the sides as the channel deepened. Trees overhung the river's banks; their roots, tannic, darkened the water with acid. We carried long-handled shovels that we used to tap the riverbed in front of us. It was slow going.

We were looking for fossils. The Bone Valley Member of the Hawthorn Formation encompasses parts of five counties—Polk, Hardee, Manatee, Hillsborough, and DeSoto—and throughout those counties are preserved a rich array of vertebrate species that originated in a time when Florida was a site of interchange within the world. Bone Valley contains the Peace River nearly from top to bottom. Its mines have provided the lion's share of the phosphate fertilizer consumed by American farms over the past century and a substantial chunk of the fertilizer used around the world. It was named by prospectors 140 years ago as they searched for phosphate by following traces of fossils in the ground.

As the work of disparate prospectors became a system of mass extraction, the phosphate industry changed its course. During the twentieth century, the pickaxes and shovels of England's mines would be replaced by towering American machinery that churned through acres every day. The Peace would be the nucleus of this new industrial organism. Fossils that tumbled downstream—so many relics of former lives—would guide humanity in a new direction. The lives that made up phosphate had been born, had grown, had reproduced; they reached their peaks and died. Their offspring had survived, and the chain continued onward. Their life-force was preserved within the rock. It was unearthed. Phosphate flooded the systems of our planet. The phosphorus cycle snapped, and the chain of life was changed as a result.

There has been, throughout the course of history, a general trend: the manner in which phosphate is extracted has often come to define the style of food production that results. Before the mining of phosphate rock, when fertilizer came by means of livestock and the cycles of the land, farming was connected to ecology. The system that developed after Henslow's walk at Bawdsey was small but scientific, just like the mining itself. When Florida became the world's dominant source of phosphate fertilizer, the features of agriculture would change again. Farming would become a larger and more corporate enterprise. Its basis would be not the cycles of nature but the notions of economics, abstract theories gaining precedence over the physical realities of the landscape. This new system would be driven by the interests of oil companies and global grain traders. Around the world, these corporations would help create a monolithic system of agriculture driven by profits rather than need. An agricultural system built on sprawling mines would be extractive from beginning to end. It would extract nutrients from soils, wealth from consumers, and memories from the ground. The defining characteristic of the new system would be a single feature: scale. This would be a massive undertaking.

The fossils of Bone Valley were once encased in a sunken expanse of phosphate rock. The sea of stone that held those bones changed the face of human existence as we know it. Mined, processed, and shipped around the globe, applied on farms, used to grow food to feed the richest countries of the world—phosphate is driving a wave of life that has yet to crest.

Bone Valley stretches out across two thousand square miles, a corridor of fragmented geology that cuts through Central Florida between Tampa and Orlando. Mining in the region today provides a quarter of the world's phosphate fertilizer. One hundred and fifty thousand acres have been stripped. Fossils, uplifted by mining, have illuminated the history of the earth. The same conditions that created phosphate—the shallow seas, the masses of dead life—also allowed for the preservation of fossils, the cell-by-cell replacement of life by facsimile mineral matter. Searching for something in the ground there—a leg, a tooth, a jawbone—is art informed by science. You look for black gravel that signifies the kind of material that might hold fossils. "It's the old rock from the seabed. Little smoothed pebbles—that's one of the indicators, if you will, when you're looking. If you don't see that, then there's no point in looking. But if you start seeing real smooth, water-smoothed pebbles, and they're black, then you know you're onto something," Thom Downs—who was then the president of the Bone Valley Gem, Mineral, and Fossil Society—said. On land, that search can be as simple as looking around. In the water, where seeing is difficult, the process is more hit-or-miss. But the water also provides more variety. It moves rocks around, creating high concentrations of fossils that have been eroded from the earth over time. "It's kind of nice, a little unnerving, being in water that you can't see the bottom of. When you're waist deep, that can be a little frightening," Downs said. "It's interesting that, however many millions of years ago, there were these other things that lived here. And I think that's pretty cool."

When Downs discards unwanted rocks from his sieve, he tosses them behind his back. He prefers to move methodically, spending hours at a certain spot, combing slowly through the river. "Jumping around you'll find some stuff, but you won't find a lot," he said as we waded upstream. "The odds get worse depending on how erratic you're doing it." He has brought home enough fossils, he told me, that he possesses "a pile of bones in the backyard."

We could not see the bottom of the river, but we listened for the crunch of fossils and phosphate beneath our feet, and we felt for holes that could trip us up. Downs paused frequently, jammed his shovel into the earth, scooped a pile of rocks, dumped them into the sieve he carried, and shook the mud. His fingers picked through what remained: sharks' teeth, manatee bones, the perfectly preserved dermal plate—an upper tailpiece—of a prehistoric alligator. He gave me the better sieve, a small plastic beach toy that was easier to use than his bulky wooden contraption, and I tried to repeat his motions on my own. His finds exceeded mine. He gave most of what he found to me.

After heading south along the riverbed, we retraced our steps northward and continued on to a narrow channel that diverted, for several hundred feet, away from the river. The channel was tight and swampy. Where fallen trees blocked our trek, we had to scurry up the edge of the bank, clamber through the mud, and slide down the grassy incline to return to the water. We lived in fear of falling. "Are you a Star Wars fan?" Downs asked me as we summited one particularly muddy peak. "This reminds me of Yoda's planet: swampy and gross."

. . .

For most of its history, Florida was covered by the water of various sloshing seas. The peninsula that exists today began its assembly half a billion years ago, when all the world's continents were connected as one. Africa and North America bordered each other, and volcanoes rose between the two, forming the base of the state as it now stands. Marine organisms, living and dying on the underwater shelf that had emerged, layered their material above volcanic rock.

One hundred and eighty million years ago, the continents of the world began to split. Africa and North America separated at a rate of one inch per year. Thirty-four million years ago, as the continents approached their present positions, cooling global temperatures caused sea levels to drop, and the Florida peninsula first began its rise from the sea. Ice formed in the polar regions of the earth. "From about fourteen million years ago, the East Antarctic Ice Sheet became a permanent feature on earth, and is still, of course, there. So that led to significant changes in global climate, led to significant changes in ocean circulation, and, as a consequence, upwelling. And the upwelling then led to this large phosphate deposit," John Compton, a geologist whose 1993 paper described the formation of the deposits, told me. The ice sheet took water from the sea, and the level of the oceans dropped. Most of the Florida peninsula remained underwater, but the water was shallower, and the creatures in it changed. Larger whales left for deeper seas, and smaller sharks and sea cows moved in. The land that emerged was populated by tortoises, miniature horses, camels, and bear dogs that congregated in a mixture of prairie and lagoon environments. In these surroundings, phosphate began to form. Among the shallow waters, creatures lived and died, depositing nutrients that would become phosphate rock.

As time passed, more of modern Florida emerged from the sea. Five million years ago, as the Miocene became the Pliocene, Florida's land areas were dominant and plentiful, with saber-toothed cats, rhinoceroses, and elephant-like animals roaming the peninsula. Then the earth cooled more. The Great American Biotic Interchange, made possible by a land bridge in Panama, introduced new life from South America into Florida and the rest of North America: armadillos, glyptodonts, mastodons, giant ground sloths, and carnivorous terror birds, which stood nine feet tall and dominated their surroundings.

During the world's last ice age, which began 2 million years ago, in the Pleistocene epoch, sea levels dropped even further, biological productivity decreased, and more of Florida's water environments became land. The phosphate rock—and the fossils it enclosed—pushed upward but remained hidden beneath layers of rock, sand, and soil. Above those layers, mastodons and mammoths thrived. Paleo-Americans arrived in Florida about 14,000 years ago and hunted its animals, driving many large mammal species to extinction. By the time Europeans came, 13,500 years later, the peninsula's ecosystems had come to resemble what they are today.

The history of superphosphate in America began in 1843, one year after Lawes and Liebig, working independently, had invented it in Europe. Despite only a scattered history of fertilizer use, Americans quickly came around to the idea of phosphate fertilizer, thanks largely to the work of the English chemist Humphry Davy. A wide-ranging researcher of electricity and other subjects, Davy was a scientific celebrity who focused for much of his life on the new field of electromagnetism. He explained the nature of electrolysis, refined the scientific understanding of acids, and discovered seven new chemical elements. He also mentored Michael Faraday, whose work would provide the foundation of modern concepts of electricity; Davy gave Faraday his first university appointment. Noted throughout Europe for his pioneering advances that used science to practical effect—including the invention of the Davy lamp, which allowed miners to safely carry light underground—Davy's popular reach in the early nineteenth century was extensive. A poet as well as a scientist, Davy counted among his friends Samuel Taylor Coleridge and Robert Southey, on whom he experimented with laughing gas, to lasting notoriety. In 1802, he began a series of lectures for the British Board of Agriculture that were published around the English-speaking world, beginning in 1813, as *Elements of Agricultural Chemistry*.

With *Elements,* Davy showed farmers how to nurture land rather than abandon it, how to apply fertilizers, and what to do to fix a sick soil. He wrote clearly and lyrically about the upkeep of the soil. "Are any of the salts of iron present?" he asked. "They may be decomposed by lime. Is there an excess of siliceous sand? The system of improvement must depend on the application of clay and calcareous matter. Is there a defect of calcareous matter? The remedy is obvious." On the subject of phosphate, he recommended "digesting together bone ashes and oil of vitriol, and strongly heating the fluid substance so produced with powdered charcoal." He offered clear advice on the application of fertilizers and the treatment of the ground: "Is an excess of vegetable matter indicated? It may be removed by liming, paring, and burning. Is there a deficiency of vegetable matter? It is to be supplied by manure."

Davy's ideas arrived like a shock in the United States, where soil health had long been generally ignored. White Americans often depleted soil and then abandoned it in favor of farming in another place. "The improvident use of originally fertile soil year after year, for cropping with grain, for example, without attempting to replace by fertilisation the plant foods which are annually taken up by the crops, is simply a method of living on capital as farmers in the Middle West of America have found to their bitter cost and now the problem of the re-creation of fertility has to be faced and it is a very long and difficult business," A. N. Gray wrote in 1944. George Washington, prominent not just for politics but also as a landowner and agriculturalist who owned and brutalized hundreds of people over the course of his lifetime, had experimented with manuring fields, as had various other people throughout the country. Thomas Jefferson, another slaveholder-politician, had focused on the effects of crop rotations as a means of improving the soil. Most mainstream farmers, however, had rarely evinced interest in protecting the health of the land. "Farmers seldom hauled manure to the fields even when they had a sufficient livestock population to sustain the productivity of the soil," the Tennessee Valley Authority's Lewis B. Nelson wrote in 1966 in the TVA-published *History of the U.S. Fertilizer Industry*. "Numerous observers reported that farmers in the Northeast would let manure accumulate for years in the barnyard, even to the point of damaging farm buildings. Eventually the farmer had to either haul out the manure or build another barn." Davy's book, brought over from England, was republished in Pennsylvania, New York, Virginia, and Connecticut and was praised lavishly by readers, creating a space for greater technological shifts in American farming. "From the time when the first shipments of the book reached these shores until the 1840's, farmers in all sections of the United States bought copies, learned about the chemical foundation of agricultural operations, and came to look upon its author as their master," the historian Wyndham Miles wrote in 1961.

Justus von Liebig idolized Davy. "Since the time of the immortal author of the *Agricultural Chemistry,* no chemist has occupied himself in studying the applications of chemical principles to the growth of vegetables and to organic processes," Liebig would write in the introduction to his own work on agricultural chemistry in 1840. "I have endeavored to follow the path marked out by Sir Humphry Davy, who based his conclusions only on that which was capable of inquiry and proof." Liebig's rise to prominence—followed closely by that of Lawes—ended Davy's supremacy in the field of agricultural chemistry. The 1842 invention of superphosphate marked a turning point in this process. It represented the end of Davy's reign and the beginning of a new, even more direct approach to agriculture that would spread throughout Europe, America, and the rest of the world.

For ten years after its invention, superphosphate was consumed, but not produced, in America. That changed in 1852 when two American businessmen, James J. Mapes and C. B. De Burg, began to manufacture and sell mixed fertilizers that included superphosphates. One year later, an American named Samuel Johnson embarked on a mission to bring not just superphosphate but the whole European agricultural research apparatus to the United States.

Johnson was an early adherent of agricultural science. Born in 1830, he was intent on replicating the Rothamsted Experimental Station in America. In 1853, he traveled to Europe to view Lawes's and Liebig's work firsthand. "Having visited the experimental farms of Lawes and Gilbert and other places of agricultural interest, he hurried back," the National Academy of Sciences would later report in a biographical sketch of Johnson. In America, Johnson began teaching analytical chemistry and later agricultural science at Yale University in New Haven. He published the book *How Crops Grow* in 1868 and followed it shortly thereafter with another volume, *How Crops Feed,* the two books serving as a new foundation of fertilizer knowledge in America.

Johnson believed that superphosphate and other innovations could make agriculture more profitable, improving the lot of farmers. He translated Liebig's chemical treatises, but he pushed beyond what Liebig had written. "The mere chemist is not fully adapted to advance the science of agriculture; for although chemistry, more than any other one department of science, may promise profit to the farmer, yet it alone is not sufficient," Johnson wrote in *The Country Gentleman* in 1853. "When a Liebig writes on the 'Relations of Chemistry to Agriculture and Physiology,' he does not compile a receipt-book, nor collect from reading, travel and experience, an account of the various improved methods of conducting farm operations. By no means. He is no farmer," he wrote three years later. Johnson wanted to create an American version of Rothamsted, but with profit at its center. "He was the apostle of a cause, and his earnestness, sincerity, and clearness as a lecturer rendered him and his subject popular among his hearers and secured for him the confidence of the agricultural community," the NAS wrote.

In 1875—following decades of lectures, article writing, and teaching—Johnson convinced the State of Connecticut to establish America's first experimental station, a small institution dedicated mainly to rooting out fraudulent fertilizers. The Connecticut station was a success, and, over the coming decades, experimental stations were built in most U.S. states. "The whole system of agricultural experiment stations may well be regarded as his monument," Yale's president would later write. The development of these facilities tracked closely with the increasingly widespread academic study of agriculture, initiated by the Morrill Act, which in 1862 had established land grant institutions where agricultural practices would be honed and studied. One result of these changes was a sharp increase in superphosphate use in America. Between 1850 and 1920, U.S.

consumption of phosphate fertilizers increased 165-fold—from 4,000 to 660,000 tons per year. Some superphosphate arrived in America from Europe—including shipments from Packard's, in Bramford, in 1877—but the United States was mostly responsible for its own production.

American use of superphosphate rose substantially, but the rarity of phosphate presented a frustrating obstacle to farmers. Just as Britain and other European powers had picked bones off battlefields and pulled fossils from rivers, producers in America collected phosphate from macabre and surprising sources. Human waste, termed *poudrette,* was harvested from cities and sold to farmers, often at exorbitant prices. Slaughterhouse waste, called *tankage,* was consumed with fervor by the fertilizer industry. Hooves were ground into meal, organs were dried, and heads were squeezed and ground into powder. Blood, which contains high levels of nitrogen, was dried and shipped around the country. Manure, fish, cottonseed meal, and bones were made into commercial fertilizers.

Fertilizer manufacturers wanted cheaper and more plentiful sources of phosphate. Europeans had probed their land looking for phosphate rock, and Americans began to do the same. In 1867, in South Carolina, the first deposits of rock phosphate in America were found, and phosphate production was transformed. The South Carolina reserves were limited, however, and geologists scoured the country, looking for a larger supply. Twenty years later, Florida filled the void.

The first phosphate rock found in Bone Valley was uncovered during a river-dredging project in 1881, but the businessmen who were alerted failed to grasp what the discovery meant. Moneyed locals on a hunting trip five years later grasped both the rock and its significance, and they quickly bought up thousands of acres of land in the vicinity. Their intention was to mine the phosphate pebbles that littered the river's bed. They commenced mining in 1888. More phosphate was soon found on dry ground nearby, and land pebble mining began two years later. It was a gold rush, an oil boom—a promise of prosperity to anyone near. "Speculating upon the wealth to be derived from these immense deposits that vary from twenty feet to unfathomable depths, is beyond our capability, it is like the enchanted Isle of Monte Christo or the wealth of Croesus—phosphate, phosphate, thou art a treasure," the *Polk County News* reported at the time. "Ere long millions upon millions will roll into our coffers and we will be the hub as it were, around which all the other states and countries will revolve."

Across America in the 1930s, phosphate was growing in importance. Unsustainable agricultural practices had destroyed large amounts of farmland, taking phosphorus from the soil and rendering vast acreages unusable. Left fallow, those phosphorus-deficient stretches of land helped to cause the Dust Bowl, a six-year epidemic of devastating dryness that descended across the Depression-era Great Plains. In 1938, in the wake of that catastrophe, Franklin D. Roosevelt decided that phosphorus was the answer to America's agricultural and environmental woes. In May 1938, Roosevelt addressed Congress on the subject. "As a result of the studies and tests of modern science, it has come to be recognized that phosphorus is a necessary element in human, in animal, and in plant nutrition. The phosphorus content of our land, following generations of cultivation, has greatly diminished. It needs replenishing," Roosevelt said. "I cannot overemphasize the importance of phosphorus not only to agriculture and soil conservation but also to the physical health and economic security of the people of the Nation."

Central to Roosevelt's speech was a concern that the United States would soon run out of phosphate. In Florida, he pointed out, more than one-third of all phosphate produced was going overseas. Since Florida's mines produced nearly three-quarters of all U.S. phosphate, that meant that a quarter of all domestic rock departed the country, even as American farmland went without needed fertilizer. Moreover, official estimates showed that the reserves in Florida would dwindle shortly, and Roosevelt worried that the country would be left with troubled land and no recourse. "It is, therefore, high time for the Nation to adopt a national policy for the production and conservation of phosphates for the benefit of this and coming generations," Roosevelt told America. In response, Congress formed the Joint Committee to Investigate the Adequacy and Use of Phosphate Resources of the United States. They wanted to create a national policy to manage the consumption of phosphorus. In order to determine how much phosphate the United States possessed, the committee scheduled hearings near mines in Idaho, Alabama, Tennessee, and Florida. At a hearing in Washington, D.C., members of the committee suggested that steps be taken to preserve what phosphate remained, including banning exports.

Proposals to curtail the phosphate export industry alarmed the fourteen companies that mined the Bone Valley region at the time. Producers in Bone Valley were sending three hundred thousand tons annually to Nazi Germany and well over two hundred thousand tons each year to Imperial Japan, lucrative sales that generated high profits. These companies were the only phosphate exporters of any magnitude in the country. To avoid an export ban, they needed to show that Florida had access to far more phosphate than U.S. farmers could possibly ever need—so all fourteen companies decided to collaborate. "With the threat of possible loss of their export trade," the geologist George R. Mansfield would later write, "they put aside for the time being considerations induced by the highly competitive nature of the business and made available for presentation to the Committee information that hitherto had been closely guarded." The companies sent a man named Wayne Thomas to testify at the Florida congressional hearing, which took place that November in Lakeland, in Polk County. In his testimony, Thomas, a prospector, told a story that directly contradicted Roosevelt's premise that America was facing a looming phosphate shortfall. Thomas told the story of his life. He spoke of decades' worth of prospecting trips around Florida, of holes he had dug, phosphate he had found, documents he had gathered, and calculations he and others had performed. To support his story, Thomas brought with him proprietary maps and data that had been created by the mining companies and never before publicly shared. It was a monumental leap of faith for the companies to pool together such carefully guarded information, placing it in the public record.

For the members of the committee, this information was unprecedented in its scope and scale. When Thomas began to estimate the quantity of one untapped reserve in Florida, Charles Leavy, a congressman from Washington, interjected.

"And that would reach figures you would hesitate to give, and that would be staggering?" he asked.

"Yes, sir," Thomas replied.

"Do you mean by that they would run into the billions?" asked Leavy.

"Many billions," Thomas said.

"That is a matter of proof?" asked James Pope, the Idaho senator who chaired the committee.

"I would assume so," Thomas said.

By the end of the testimony, Thomas's estimates of Florida's phosphate resources had soared far above the estimate given by Roosevelt in his address to Congress—500 million tons—as well as all earlier numbers that had been publicly shared by the phosphate companies.

"A few feet of the top crust," Thomas said at one point, "will yield twenty billion tons."

Claude Pepper, U.S. senator from Florida, spoke up, incredulous. "Did you say twenty billion?" he asked.

"Yes, sir," Thomas said. "It may be forty billion."

The committee was dumbfounded by Thomas's testimony, which proposed the presence of far more phosphate resources—possibly a hundred-fold more—than had previously been publicly known to exist in Florida. These additional reserves, if they were real, could add thousands of years to the longevity of America's phosphate industry. Thomas, along with other representatives of the Florida producers, argued that a national system of phosphorus conservation would hurt, not help, the nation's farmers. Their argument prevailed, and the companies' phosphate exports continued.

In the end, both Roosevelt and Thomas were correct. American soil did need phosphorus—but the contents of the Florida reserves were more in line with Thomas's estimates than with Roosevelt's. The mining could continue unabated without the rock running out—but months after the hearings ended, World War II began, and phosphate exports to Germany, Japan, and elsewhere mostly ceased. Then the war ended, the export market grew again, and the modern Florida phosphate mining operation began.

The day we searched for fossils, Thom Downs wore camo-patterned cargo pants, tattered Sperry sneakers, and a long-sleeved sun protection shirt with a collar and a vent. We met in Lakeland, where Roosevelt's hearings took place nearly a hundred years earlier, and we drove an hour south to Hardee County, where 12 percent of land has now been mined for phosphate. Hardee straddles the Peace River. We entered the river near the little city of Bowling Green, on the northern edge of Hardee, and that was where we found the dermal plate, where we climbed through Yoda's planet, and where a fish—we hoped—splashed our faces. We drove south.

At Gardner, near the southern edge of Hardee, we picked our way along the side of the Peace, sticking our shovels into the riverbed and occasionally scooping gravel. Above us, the banks rose steeply; before us, the river extended far and curved left. The channel here was wider and deeper than before. Downs was warier. There were bones beneath us, teeth and shells. The sun was out. We climbed a bluff, and we looked out at the river in all its curving majesty below. My pockets were filled with phosphate pebbles the miners never got.

In the 1980s, when Downs was growing up in Polk County, the mines were excavating millions of tons of phosphate each year, raising history to the surface. Thousands of acres at a time were clear-cut and strip-mined fifty feet deep. Linear trenches dissected the area, running alongside larger pits and veritable mountains of rubble. The pits were deep enough to break through the water table, and giant pumps removed water day and night. Changes in the landscape were wrought by draglines, huge machines with dangling scoops, each scoop large enough to fit a truck inside. The draglines, which were too heavy to be transported, walked on their own between locations. The landscape they traversed was one they had created, a sea of white littered with holes, the horizon punctuated by machinery.

In those days, the companies that dominated the industry—IMC, Agrico Chemicals, Conserv, Cargill, and others—served as the benevolent gatekeepers of Florida paleontology. Mine managers worked out arrangements with local people who wanted to find the fossils that were being excavated. "We had a relationship with one of the guys at the security gate—you know, names were on the list. You could go in," Downs said. "You could walk right up to the edge and look down. You'd see them with the dragline, scooping stuff up and dumping it—the big hoses blasting everything into slurry, which they'd pump out. It just went on forever." Fossils were everywhere: dumped into piles, exposed along the cut, strewn haphazardly across the ground. "I remember one time a guy climbed up out of the pit, and he was carrying these—I don't know how he did it, but he had a bag full of limestone with really, really big sand dollars fossilized," Downs said. "That is insane, thinking back. He was climbing out, and I was like, holy hell."

Over time, however, outside firms began buying property. Oil companies invested heavily in the Bone Valley mines. ("The oil industry has the capital, research capabilities, and international structure essential to the phosphate industry today," the historian Arch Fredric Blakey had written in 1973.) The mining companies consolidated. In 2004, the last two major corporations, Cargill Crop Nutrition and IMC Global, merged to form the Mosaic Company, a Minnesota-based conglomerate that became the largest integrated phosphate mining company in the world. Mosaic acquired mines operated by Florida's last remaining independent phosphate mining company, CF Industries, in 2014. In 2018, Mosaic relocated to Tampa. Amid this consolidation, access to the mines was restricted and the future of fossil-hunting was clear. "It's not an option now," Thom Downs said. Fred Hendershot, who organizes trips for the Tampa Bay Fossil Club, told me, "They just shut us off completely." No more fossilists would climb the cuts and overburden piles; no more bones would be preserved. The mines themselves would grow, but their contents would be treated indiscriminately. The record would be excavated, the fossils dug out, crushed, and then compacted into fertilizers, the pages of history burned to fuel the lives we lead today.

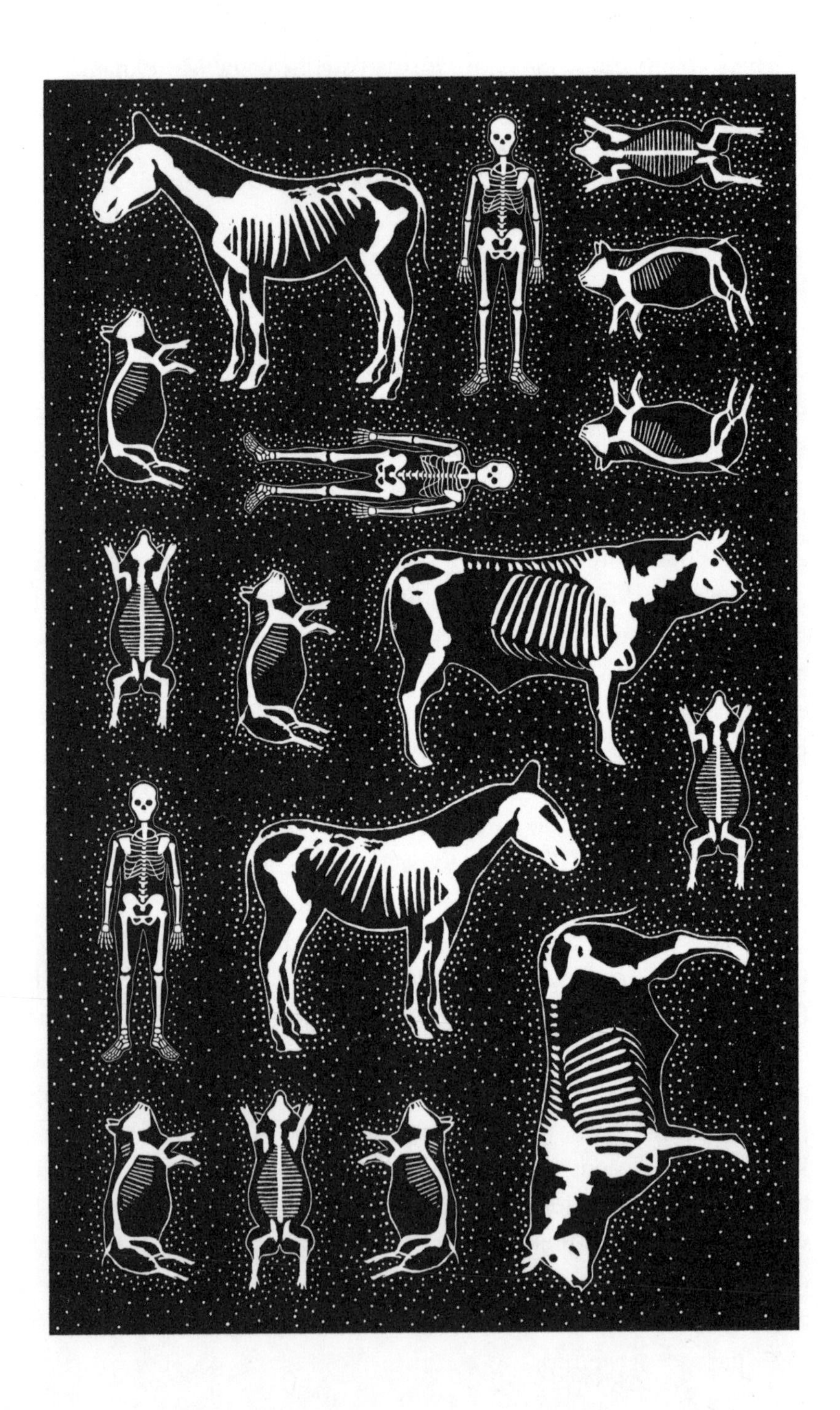

5

RAPID CHANGE

A white phosphorus projectile is a munition that, when released into the air, flies a distance and then breaks into innumerable little flaming bits that rain down from the pinnacle, spreading out to form a field of burning light. From a distance, the explosion looks like a burst of languid fireworks. Up close, it burns to the bone. The purpose of white phosphorus, it has been said, is "to camouflage movement." It also can be used to illuminate an enemy: the light burns ever brightly in the dark. Officially, white phosphorus should never fall on people. As long as the munitions are not used on human bodies, they are permitted by international humanitarian law. But the fire creates a smoke that hides the truth. White phosphorus falls on people all the time.

It was used by the British in World War I and later against colonial rebels in Iraq. Allied troops bombarded German soldiers with white phosphorus throughout World War II. U.S. forces used it in Vietnam, the Russians in the Chechen Wars, and the British when they took back the Falklands. At Fallujah, in Iraq, U.S. forces bombed their enemies with white phosphorus, and in Afghanistan, Taliban militants fired white phosphorus at U.S. troops. Israeli forces have fired white phosphorus into Gaza and Lebanon, while the government of Syria has used it repeatedly. The Russian military has been accused of shooting white phosphorus at civilians in Ukraine. In some of these cases, the use of phosphorus has been denied by the governments under accusation of using it. Other times, governments have claimed they used it for illumination or cover. Behind the screen of fire and smoke, however, there were charred limbs and melting skin, singed hair, dead bodies, and destroyed homes. White phosphorus burns without mercy, and the damage that it leaves is breathtaking.

A phosphate molecule is made of phosphorus and oxygen bound together. To make a bomb, all of the oxygen is stripped away, leaving only the phosphorus atom behind. When phosphorus combusts—when a weapon explodes or a match ignites—it combines again with oxygen, returning the phosphate molecule to existence. The manufacture of these bombs was pioneered during World War I and reached its zenith during World War II. Factories turned out phosphorus and nitrogen munitions throughout the United States and the world. When the wars ended, those factories emptied, and a new campaign began to sell phosphorus and nitrogen as fertilizer. This campaign was conducted by some of the world's wealthiest businessmen and some of its most powerful governments, and it changed the way agriculture was practiced. Phosphorus munitions propelled the supply of fertilizer, and phosphate became an industry on which a world of people relied.

It is a peculiarity of phosphorus: with a single substance, humans grow and humans kill. As extraction of phosphate rock expanded, so too did new consequences arise. Included in the phosphate crystal structure are various heavy metals and radioactive elements that come to the surface when the rock is mined. As the rock is processed into fertilizer, the land, air, and water all present potential pathways for these dangerous elements to spread. As toxic waste rushed down rivers and acid spewed into the skies, mining companies worked to limit the damage they had caused. But problems grew, ecosystems suffered, and people reported illnesses near the mining sites. Once phosphate is taken from the ground, the consequences are difficult to control.

In 1964, soon after they were married, Norma and John Killebrew decided to buy a piece of land, build a house, and start a life together.

They chose an area of southeast Hillsborough County, Florida, in a corner that borders Polk, Hardee, and Manatee Counties. They purchased 49.5 acres of farmland and a herd of cattle. The land around the Killebrews' home was a sea of ranches divided by country roads. The Little Manatee River, which eventually flows into Tampa Bay, found its headwaters in the swamps behind where their cattle grazed. The pastures were fertile and grassy. Their home was breezy. Norma taught English at an inner-city high school, and John took a job as an ironworker in Polk County. They had three children.

In 1972, the Killebrews received a flyer in the mail from the Army Corps of Engineers inviting them to attend a meeting in Pinecrest, a community thirteen miles to the north. The subject of the meeting was phosphate: IMC, a company that operated phosphate mines in Florida, had asked the county to rezone land to allow for mining near the Killebrews' ranch. "We attended the meeting because we had an investment in the property," Norma Killebrew told me. "And then we were told that our area would be rezoned, and we found out. We were ignorant. We weren't really smart back in those days." After that first meeting, the planning process seemed to move rapidly forward. There were more meetings, many of them far away from the Killebrews' house. "We didn't have the luxury of coming in and being able to drive thirty, forty minutes into town when there were meetings at the downtown office, the Hillsborough County Commissioners office," she said.

By the time those meetings finished, Hillsborough County had rezoned 53,000 acres of land to allow for mining. In all, the county approved three separate mines that bordered one another: Lonesome in 1974, Kingsford in 1975, and Four Corners in 1978. The rezoned land surrounded the Killebrews' ranch. "It was all woods. There was viable grass. You could put cattle on it. There had been mound Indians in the area—there was an archaeological dig up the road, if you take a left at the general store. The property across the road had cattle on it," Norma Killebrew said. After IMC gained mining approvals, it leased the land to agribusinesses while the timeline for the mines continued. Fourteen years after the initial meeting in Pinecrest, mining began. Over the decades that followed, the Killebrews found themselves surrounded on all sides by miles of white. The operation, known today as Four Corners, would become the largest phosphate mine in the world. The industry, long based in Polk County at the northern end of Bone Valley, was growing.

In the decades before Four Corners was constructed, American agriculture had undergone a rapid transformation that resulted in higher national consumption of phosphate rock. America's farms had long been small, local, and diversified. Since the eighteenth century, farmers had raised animals and food crops, spreading manure on the ground and using phosphate fertilizers when not enough manure was available. They had managed their own affairs by paying careful attention to the environments in which they lived and the communities that had allowed them to prosper. Technological advances—and the growing use of superphosphate—had made agriculture more scientific than before, but in the early twentieth century, farms still operated based on local knowledge and local need.

At the time, economists were grappling with overarching questions about the creation of wealth and the basis of growth. During World War II, the government had taken an active role in everyday affairs by regulating industry, providing price supports to farmers, and constructing a social safety net to protect the poor. The advent of a more controlled economy—coupled with the rise of communist systems abroad—forced a series of reconsiderations of work and wealth in America. Some economists retained a long-held belief that natural resources, including agriculture, were the main drivers of growth. Others began to see banks as more important economic actors; the financial sector, they felt, could be trusted to organize the market appropriately. Franklin D. Roosevelt had held up small farmers as crucial to ideals of economic development. Under Roosevelt's policies, minimum prices for agricultural goods were guaranteed, insulating farmers from market forces. Harry Truman had maintained these policies, and his reelection as president in 1948 was supported by a populist alliance of farmers and urban workers intent on holding on to the gains they had made over the past twenty years. During Dwight D. Eisenhower's administration, however, those policies were challenged. The new secretary of agriculture, Ezra Taft Benson, stocked Eisenhower's agriculture department with businessmen who advocated for the consolidation of American farms. These new policymakers opposed the price supports that ensured the stability of small farmers, advocating instead for corporate subsidies that would increase production of cheap grain. Earl Butz, an acolyte of Benson's who would go on to serve as agriculture secretary, told farmers to "get big or get out." The pendulum of policy swung decidedly toward agribusiness.

Under the direction of Benson, Butz, and many others in business, government, and academia, family farmers changed their ways. They turned small farms into booming factories, borrowing heavily to do so. They bought fertilizers, pesticides, antibiotics, specialized equipment, and many other expensive goods from the companies that preached the benefits of industrialized agriculture. They bought their neighbors' land, often turning multiple interdependent family farms into a single operation that produced just one crop. The changes advocated by the business-led Committee for Economic Development, the American Farm Bureau Federation, the U.S. Chamber of Commerce, and a host of agribusinesses—including Cargill, which also played a major role in phosphate mining in Florida—brought about a greater reliance on phosphate fertilizers in the United States. The organizations that advocated for this new system believed they had found a cure for food insecurity, and they sought to share their answer with the world.

The events that would change global agriculture were underway long before Eisenhower took the presidency. The 1940s were a time of change around the world. Fears of fascism had been replaced by concerns over communism. Human populations ticked steadily upward, and some scientists were worried that food production would not keep pace. Political upheaval and food insecurity often intersected; amid hunger, politicians worried, revolution might foment. In 1940, when Roosevelt ran for reelection, he chose as vice president his longtime agriculture secretary, the Iowan farmer Henry Wallace. A month after the pair were elected, Wallace traveled to Mexico to attend the presidential inauguration of Manuel Ávila Camacho, whose election was seen as a bulwark of democracy. While in Mexico, Wallace toured peasant farms, departing with the feeling that their practices could be enhanced by different seeds. Back home, Wallace met with representatives of the Rockefeller Foundation to discuss his trip. At the time, the Rockefeller Foundation was a young organization with a stated goal of solving fundamental global issues using science. It had been created by John D. Rockefeller and funded by the wealth that resulted from the sale of Standard Oil, which had operated as a monopoly. Since the 1930s, foundation leaders had taken an interest in agricultural issues. In 1941, at the urging of Wallace, they looked toward Mexico. The foundation conducted its own survey of Mexican agriculture, sending three scientists on a five-thousand-mile journey by car. By October, with a growing collection of geneticists on a 150-acre plot of land at Mexico's National School of Agriculture, the Mexican Agricultural Program (MAP) had begun.

The Rockefeller Foundation, working to change agriculture in Mexico, at first sought to preserve local systems of farming. Despite growing consolidation in the United States, the vast majority of the world still relied on small farmers for food. Technological progress on farms came from the bottom up, not from the top down. Farms still worked like ecosystems. Realizing the importance of the Mexican *ejido* system, through which small-scale farmers grew corn for national sustenance, the leaders of MAP sought to improve corn varieties while working with the local population. The program was effective, and incremental progress was made. Throughout the 1940s, however, the program's focus changed. Norman Borlaug, a geneticist working at a MAP facility seven hundred miles to the north, successfully pushed for the creation of a new program focused on wheat. This was a significant shift: whereas corn was grown by small farmers to feed the general populace, wheat was grown on large commercial farms in the north, and it was not a staple food crop. Borlaug saw this contrast as an opportunity. He wanted to transform the entire system. "In 'traditional' agriculture, crops were bred to survive in meager soils; sparse plantings offered defense against pests and diseases. Cultivators adjusted their aspirations to what the land could provide.

Raising yields required improving the soil but also developing varieties that could take the higher fertilizer load," Nick Cullather wrote in a history of the program. Borlaug "came to see modernization as a transition from lower to higher levels of soil nutrients." Working with corporate growers in the north, Borlaug designed new wheat seeds, but the MAP's corn program languished in the south. In 1948, before much had actually been accomplished, Mexico experienced a successful harvest, and the Rockefeller Foundation took credit for it. The foundation launched a public relations campaign claiming to have solved Mexico's food production problems. It said that Mexico could be a blueprint for the world.

The Rockefeller Foundation and the U.S. government next turned their attention to India, a place where revolution was fomenting, the population was growing, and the bonds that held society together seemed ready to fray. India was a country of small farmers who labored under the jurisdiction of wealthy, often absentee landlords. Land reform seemed to be the answer, but the solution Rockefeller turned to was technology: machines, genetic engineering, and chemical fertilizers. Larger farms—corporate entities with extensive irrigation systems and money available for investments—were prioritized in the rush of technological change.

In 1960, as reforms progressed in India, the Rockefeller and Ford Foundations joined together to found the International Rice Research Institute, which sought to expand the scale of the Mexican and Indian projects. The IRRI was set up at an extravagant modernist headquarters in the Philippines. Geneticists working there developed new seeds, and agronomists trained locals in the techniques of mechanical agriculture. In the face of unrest across Asia, the IRRI's goal was to bring development and prosperity to the masses of people tempted by communism. Its leaders wanted to remake Asia.

One by one, geneticists raised the limiting factors on crops. The development of the photoperiod-insensitivity gene allowed plants to grow year-round, not just during a particular season. Rice in Southeast Asia was bred to mature in only 105 days, rather than the norm of 170. Varieties were developed that could withstand longer floods, more acidic soils, and various kinds of natural toxicity. Corn plants were made to grow closer together. Species were selected to better withstand drought. Wheat, which grew tall and spindly, was engineered to grow shorter, making it more compact and also more efficient, because it produced less straw. The shapes of leaves were changed: they drooped less in order to absorb more sun. Seed diversity declined. "A relatively small number of varieties replaced literally thousands of traditional varieties," Gurdev S. Khush, a geneticist who spent thirty years working with the IRRI to develop higher-yielding varieties, wrote in *Nature* in 2001.

The wheat seeds were planted in India, the IRRI rice was planted throughout Southeast Asia, new corn strands were growing in Mexico—in the spring of 1968, global yields surged, grain prices plummeted, and William Gaud, the director of the U.S. Agency for International Development, gave a speech that called this movement the "green revolution." The harvest was striking: so much grain that bags were left to rot in India, where there were anecdotal reports of harvests five times the size of those from the previous year. There were regional yield increases of nearly 50 percent in Pakistan. There was enough rice in Vietnam to calm the country's restive state. As the green revolution progressed, the changes were significant. Between 1950 and 1978, wheat and rice production increased 1.5 times worldwide, and corn increased by even more than that.

The growth in yields—the measure of output per area of land—was a significant component of these changes. Wheat yields increased 83 percent between 1950 and 1978. Rice yields went up 60 percent, corn yields 99 percent. Around the world, agriculture had changed. Disease had been managed, time had been rewritten, and space had been made inconsequential. Wells had been dug to irrigate the new crops, which consumed disproportionate amounts of water. Machines had been built to manage large tracts of land. Food systems had become more global as farmers became more reliant on fertilizer inputs. Half of global agricultural yield increases from the green revolution were attributed to the expanded use of synthetic fertilizers. One kilogram of fertilizer added to one hectare of land could be expected to produce an additional ten kilograms of grain.

The fertilizers were made, generally speaking, of two elements: nitrogen and phosphorus. While both are essential to the growth of plants, they serve different functions. A plant grows by creating new cells, and phosphorus atoms are needed to construct cell membranes, create new DNA, and produce the energy needed to complete these processes. Nitrogen, too, plays an important role, helping form the proteins that make up each new cell. The elements fulfill different functions, and each must be present in sufficient quantity. If nitrogen or phosphorus is missing from the soil, then a plant struggles to make new cells, and the structure of the plant suffers. The plant might lose its color, or its leaves might be deformed. The food it produces will be limited and weak. Both elements are needed, and neither can be replaced. (This was the essence of Liebig's law of the minimum, which stated that the growth of plants is limited by the rarest ingredient that they need. Phosphorus and nitrogen are just two of the most important, but least common, elements that plants absorb, and so they often are the focus of farmers' need. Another element, potassium, is also frequently included in fertilizers.)

Nitrogen is common on the earth, but much of it is contained, inert, in the air: 70 percent of the atmosphere is made of nitrogen, and it does not easily mix into the soil. The Haber-Bosch process, developed in 1909, solved this problem. Fritz Haber and Carl Bosch found a way to pull nitrogen from the air and turn it into ammonia; both men won Nobel Prizes—in different years—for their work. The Haber-Bosch process became essential to the green revolution as ammonia-based nitrogen fertilizers constituted a crucial part of the technology that was spread by scientists around the world. By the end of the twentieth century, farmers were applying 77 million tons of nitrogen each year. Today, the majority of inputs into the global nitrogen cycle come from agriculture alone.

Phosphorus, uncommon on earth, could not be extracted from the air. It had to be mined. Global production of superphosphate increased nineteen-fold between 1937 and 1981. During the 1970s, more than 70 percent of phosphate rock mined in the United States came from Bone Valley. By 1980, when 152 million tons of phosphate rock were produced around the world, one-third of the world's commercial phosphate came from Florida. It was during this time that the mines around the Killebrews' ranch were all approved. The industry needed to grow. On this larger scale, new problems became apparent.

Phosphate, like all minerals, is crystalline in nature: its structure is a latticework of atoms, a rigid frame that is often hard to break. A pure crystal consists of a homogenous array of atoms. In nature, this occurs mainly in the case of gemstones, which form under pressure. In a lab, through a process called molecular-beam epitaxy, a perfect crystal can be made; atoms are deposited slowly, in a carefully designated order. Nature lacks the careful order of a laboratory process. Phosphate forms in seawater, which contains trace amounts of many different elements. Some of those elements gravitate to the growing rock, replacing pieces and making it impure. Embedded in phosphate rock are uranium, arsenic, cadmium, chromium, and lead. When that rock is processed into fertilizer, the acid smashes the crystalline structure of the rock, rearranging its atoms and isolating phosphorus in the form of phosphoric acid, which can be absorbed by plants. The result is a more efficient fertilizer, but with a significant side effect: about half of the mass of the rock is composed of other elements, and the production of fertilizer leaves those elements behind.

One such element is fluorine, which is also an ingredient in toothpaste. Bones and teeth are made largely of apatite, a category of mineral that includes various types of phosphate rock. Fluoride ions replace some phosphorus atoms in teeth, strengthening their resistance to cavities. The fluorine in the sea caused a similar effect during the formation of phosphate rock: it replaced some ions in the phosphate crystal latticework, strengthening its structure. As the phosphate rock was mined, fluorine remained within it. In the fertilizer factories, fluorine was taken from the rock and concentrated. Highly concentrated, however, fluorine is a deadly toxin. In 1948, when Armour Agricultural Chemical Company opened Florida's first ever fertilizer plant that produced its own sulfuric acid and phosphoric acid, the effects of fluorine were widespread. "Brahma cattle began dying of a mysterious epidemic. In the middle of plentiful grass and forage, they died of malnutrition, leaving their strangely-knotted bones alongside the barbed wire that wore out every few years," one observer wrote. "In a county that by itself produces more citrus fruit than all of California, leaves were scorched, oranges grew the size of plums, and young plantings grew into middle-aged midgets." The strange effects were traced back to fluorine, which the factory was separating from phosphorus and releasing as gas. Local plants and animals were suffering from fluorosis, a crippling condition that resulted in thousands of livestock deaths.

In 1969, as part of a special report on air pollution, *Life* magazine ran a two-page photo spread of an industrial site in Florida alongside a caption that read, "Across a man-made lake, a fertilizer-producing phosphate plant lights up the night sky burning off wastes, including fluoride compounds. When the fluorides hit the ground they contaminate it, stunting and killing citrus trees and softening the bones of grazing cattle. In the last 20 years, pollution from phosphate plants in Polk and Hillsborough counties has cost citrus growers and cattlemen about fourteen million dollars in damages. People living there are reluctant to grow vegetables because of this rain of chemicals. Many of them go around all day with sore throats." Florida was not alone: In the 1950s and 1960s, as farmers in Polk County dealt with fluorosis in their crops and cattle—including an estimated thirty thousand deaths—farmers in São Paulo, Montana, and Ontario did as well. A 1967 Canadian Broadcasting Corporation report claimed that fluorine pollution from phosphate mining created an "air of death." In 1948, the same year the Armour plant opened in Polk County, a five-day smog emerged from the steel mills of Donora, Pennsylvania, killing twenty people in five days and fifty more in the month that followed—including Lukasz Musial, the father of future Cardinals baseball legend Stan Musial. Nearly half the town's fourteen thousand residents were sickened, and in the decade that followed, rates of death from respiratory illness rose. Their deaths were connected to high levels of fluorine present in the smog. The owner of the mills, American Steel & Wire, paid the victims, but it never accepted blame for their demise.

It took decades, but the phosphate industry eventually responded to these crises by installing new technology that captured the fluorine. That solution presented a new problem: massive amounts of fluorine, toxic in concentration, were being produced, and the industry needed to find a method of disposal. Dentists had been experimenting with fluoride and teeth since 1909, and in 1945 a study had begun that involved adding fluorine to the water in Grand Rapids, Michigan. The goal of the study was to see whether dental cavities in Grand Rapids would be lower than in comparable cities. The researchers found that dental cavities did decrease, but scientific comparisons were impossible to make because another city, intended to function as a control, dropped out of the study. Moreover, the fluoridation of water in Grand Rapids had coincided with other, more significant improvements in dental hygiene—namely, the widespread use of toothpaste and a standardization of toothbrushing. The science was unclear, but the dental establishment and the U.S. government pushed ahead. In 1950 the American Dental Association recommended that fluoride be added to drinking water, and the U.S. Public Health Service—a part of what is now the Department of Health and Human Services—officially recommended water fluoridation the following year.

In response to these recommendations, American municipalities sought to fluoridate their own water. They needed a source of fluoride, and they found it in Florida. The deadly fluorine that had escaped into the air of Bone Valley and then been summarily scrubbed and held as toxic waste was sold to towns and cities across America. A liability had become a profit source. Today, 200 million Americans drink artificially fluoridated water, and 90 percent of that fluoride comes from phosphate mining. It is an ambiguous tradeoff. While there is no indication that this fluoride is risky in any way—despite the protestations of various anti-fluoride advocacy groups—there is also little evidence that artificial fluoridation is particularly useful to anyone outside the phosphate industry. Water fluoridation is a clever trick for waste disposal, and a striking source of profits for Mosaic and other companies. It is "an ideal environmental solution to a long-standing problem. By recovering by-product fluosilicic acid from fertilizer manufacturing, water and air pollution are minimized, and water utilities have a low-cost source of fluoride available to them," Rebecca Hanmer, then the deputy assistant administrator for water at the U.S. Environmental Protection Agency, wrote in 1983.

As time passes and phosphate rock is formed, it also receives from the seawater around it an element that poses significant danger to all life: uranium. One ton of phosphate rock can be expected to contain somewhere in the range of one-tenth to one-half of a pound of uranium. The levels of radioactivity present in phosphate rock are somewhere in the range of sixty times higher than the amount of radioactivity typically present in the environment. In some situations, this has been fortuitous. In the 1950s and 1960s, the U.S. military's nuclear program relied on 17,150 tons of uranium supplied by phosphate mining in Florida; in the 1970s and the 1990s, 20 percent of all U.S.-produced uranium came from phosphate mining. In 1979, the U.S. Department of the Interior reported that Florida phosphate manufacturers possessed a collective annual production capacity of 3.1 million pounds of yellowcake. There is uranium in phosphate all over the world. Israel has mined low-grade phosphate ore in the Negev Desert since the 1960s to supply the country's nuclear weapons program with uranium. Egypt and Syria, which each control significant phosphate deposits, began extracting uranium from phosphate in the late 1990s. In the 1980s, when Saddam Hussein needed uranium to power Iraq's nuclear weapons development program, he hired the Belgian firm Sybetra to develop a phosphate mine at Al Qaim, near the Syrian border. Beginning in 1985, the Al Qaim Superphosphate Fertilizer Plant produced a total of 120 tons of uranium, constituting the Iraqi nuclear program's main source of fuel. In 1991, during the Gulf War, the fertilizer factory was targeted and destroyed by American forces on the first day of the invasion.

Radioactive isotopes are useful, and dangerous, because, while they are incredibly unstable, they concentrate energy to a high degree.

The movement of a particle through space can be modeled by what quantum physicists like to call an amplitude, which the rest of us typically refer to as a wave. The half-life of uranium is 4.5 billion years. As uranium atoms found in phosphate lose their protons, they decay to become, eventually, lead. When a uranium atom decays, it releases radiation, mainly in the form of photons, which move in a pattern that roughly resembles a repeating S curve. These photons interact with the cells through which they pass. They disrupt the cells' DNA, causing mutations. The mutation of a gene prevents that gene from carrying out its function. If the mutated gene is responsible for doing something small—producing water, perhaps—then the failure of the cell to perform that task will be insignificant; many other unaffected cells will continue to produce water. A problem arises, however, if the mutation affects a gene that is tasked with regulating the growth of the cell. If that gene cannot do its job, then the cell it lives in multiplies. Cancer forms.

Because of the presence of radioactivity in phosphate rock, phosphate mining has long been associated with human exposure to dangerous chemicals and the health effects that can result. Risks occur at every step of the mining process—from groundwater contamination, from acid application in factories, from waste-filled phosphogypsum stacks, and from dust and the contamination of the land. These risks affect workers, particularly in the factories, and they also affect people who breathe the air and drink the water nearby. In Israel, where Rotem Amfert, a subsidiary of Israel Chemicals (now ICL Group), estimates that 65 million tons of phosphate could be extracted from a Negev Desert deposit, a war is being waged over the health risks of extraction. For more than twenty years, people living around the town of Arad have resisted construction of the Sde Barir mine, which would displace a Palestinian village, disrupt several Bedouin towns, and uplift uranium in the vicinity of thousands of homes. "Each ton of phosphate contains tens of grams of uranium—which may be bad news for anybody breathing the dust, but is certainly a charming thought to policy makers," *Haaretz* wrote in 2016. A local man named Onn Cohen, whose father worked for another phosphate mine for forty years and later died of leukemia, told the newspaper: "When you walk around the oncology and hematology labs, it's clear as day . . . one senses hints and looks by doctors, and statements like, 'Ah, you work at Rotem.' The silence is deafening. One can't talk about the correlation and nobody does the statistics, but words aren't necessary. I once made a list of the dead. Most hadn't reached 70 and didn't enjoy their huge company pensions for a second. You see countless ICL workers in these wards, you see your neighbors fall sick—that's enough for me. I don't need anybody to tell me if it's dangerous or not. And if you do need somebody, well, the Health Ministry is saying it." Yael German, who served as Israeli health minister, had advocated against the mine following a Ministry of Health report estimating that air pollution, including uranium contamination and the release of radon gas, would lead to multiple deaths annually in the surrounding community. In 2021, the Israeli Supreme Court ruled that the mine could not proceed without additional research into the health risks it would pose.

In Idaho, whose mountains hold a major piece of the Permian age Phosphoria deposit, seventeen former phosphate mines are designated as federal Superfund sites, with the EPA assuming responsibility for managing their toxic elements. One of the sites, Gay Mine, tore up seven thousand acres of indigenous Shoshone-Bannock lands in the process of producing 100 million tons of phosphate rock during the twentieth century. The rock that was removed contained, among other elements, lead, radium, thorium, polonium-210, and selenium, which remained in the environment after the mine closed in 1993. At very low levels, selenium is a normal and necessary component of biological systems. As selenium concentration increases, however, it can cause physical pain, reproductive disruption, and neurological disorders. Monsanto, the global chemical giant, operates an Idaho phosphate mine that produces elemental phosphorus for use in the herbicide glyphosate, sold as Roundup. In 2007, the EPA investigated Monsanto for releasing selenium into local waterways—fifteen of which demonstrated elevated selenium levels—and in 2011, Monsanto paid a $1.4 million fine to settle the matter. Selenium released by the Gay Mine, the South Rasmussen Mine, and other phosphate mines across Idaho have been linked to the deaths of horses, sheep, and cattle over the past several decades. Elevated selenium levels were detected in the animals' blood. Four months after Monsanto was fined, it received a permit to construct a new phosphate mine in the state.

In 1991, Gary Pittman, whose job it was to deal with some of those elements at a superphosphate plant in North Florida, began experiencing muscle weakness. In 1993, he retired and was diagnosed with polymyositis, a potentially fatal disease. He then received a diagnosis of myopathy, a condition that affects movement, and experienced problems with his autoimmune system. Weak, dizzy, and short of breath, Pittman experienced periods of short-term memory loss. "I would be driving along, and all of a sudden, I wouldn't know where I was," he later wrote of the experience. Bedridden by the end of 1993, he had to hire a lawn service to cut his grass. He began hearing from former colleagues at the plant who were suffering from ailments similar to his. They believed that the chemicals they were putting into creeks had been poisoning their own bodies as well. In 1994, Pittman sued Occidental, which owned the plant. His body was broken, his savings had been depleted by illness, he thought that he was going to die, and he wanted the company to help support his family in case that happened. After six years of litigation, the two parties settled out of court. "They asked me what it would take to get rid of me. And there was a yellow pad sitting on this table. I took this pad and a pencil and wrote down a number. This was after we'd been to mediations ten times, but I just kept driving on. I got all their citations from the Freedom of Information Act. I would back up everything I was saying with records. I wrote a number down, and the lawyers said, 'We can do that.' And I said, 'If you can do that, I'll go away.'" Pittman said he received an amount of money in the high six figures. He is not supposed to talk about it.

"You have phosphoric acid and sulfuric acid. You've got all the metals there. You've got magnesium and manganese, you've got beryllium. . . . You've got common metals: iron, aluminum, magnesium. You've got cadmium, chromium. There's a gob of stuff in there. When they release that, all that stuff's going into the creeks," Pittman told me. "All this stuff goes into the water, and all they do to treat it is raise the pH. If it goes below seven, a light turns on and a horn starts going. That alarm tells you to dump some lime. What's that doing? It keeps you from burning the scales off the fish. They figure, once it goes in the river, it'll get diluted. And it does get diluted, to a point. But it's still there," he said.

"In the early days, I don't think the people much cared. The area was depressed. I don't think people had a wide perspective on what was going on. But as the years went on, people knew. People knew. We'd talk about it amongst ourselves. You'd be coming to work, you'd get a cup of coffee, everybody would be sitting together, and somebody might come in and say, 'You see all those dead fish out there this morning?' We'd all say yes. And someone would say, 'We must have had a big spill out there last night.' Another time someone might come in and say, 'The fumes are so bad out there today, you can barely breathe.' Or the dust would be really bad. Or somebody would be sick, he would be out. And people would ask, 'What happened to him?' And they'd say, 'Well, he cleaned the dissolver last night. I think he's got chemical pneumonia.' But all of us felt like it was our job to maintain the secrets. We felt like we were included in that team of people. We knew it wasn't good for the rivers. We knew it wasn't good for the atmosphere. But we knew we needed jobs. I started out there at eighteen. As time went on, I knew it wasn't good. I knew that they were harming the environment. We knew. They knew. They're a little smarter now about covering these things up. I don't think they talk as freely anymore. I'm not supposed to be talking now, but what am I going to do? Just sit back with this knowledge I have and let people be poisoned to death? I can't do that."

In 1960, four years before the Killebrews bought their ranch, a young couple bought a house in another part of Hillsborough County, seventeen miles to the northwest. Amos and Sarah Sheffield's home was built in a community whose existence was an experiment in upward mobility. In the late 1950s, a new interstate highway had obliterated significant affluent Black neighborhoods of Tampa, forcing residents to move elsewhere. Some people, hoping to re-create their former lives, supported the construction of a new neighborhood several miles to the southeast of the city. They called it Progress Village, and in 1960 the first residents arrived. "It was as if you were reliving the fifties rebirth of a community all over again, after the Second World War when communities were being reimagined for people moving outside of cities. Progress Village was like that, but it was a community that was built for Black people," Alfred Sheffield, Amos and Sarah's son, told me. "Progress Village was a beautiful community when we moved in." The neighborhood had its own newspaper and sports fields. In the early days, Emanuel P. Johnson, the honorary mayor of the community, "would drive around in his truck with a bullhorn attached and shout out the news of the neighborhood," Laura Baum wrote in a graduate thesis for the University of South Florida in 2016. As a child, Alfred Sheffield played in streams near his house, watched birds, and trapped turtles. He was living in a community built on contradiction, a place that elevated the stature of the Black community while also cordoning it off from the rest of the city. His friends' parents had good jobs. They lived in safe homes. But a mile and a half from the Sheffields' house, powerful acids were turning phosphate rock into phosphate fertilizer, leaving a material called phosphogypsum nearby.

When phosphate is mined, various heavy metals, including, as we know, uranium, leak into the surrounding environment, where they sometimes infiltrate the water and the land. Most of those elements, however, remain a part of the rock. At fertilizer factories, the rock is broken up with acid, and phosphorus is separated from the other elements. The main by-product of this process is phosphogypsum, a contaminated form of gypsum. Gypsum is a commonly used building material, a component of plaster, drywall, and cement. Alabaster is made of gypsum, as is plaster of paris. Gypsum occurs naturally in various kinds of sedimentary rocks, from which it has, for many years, been mined. It is soft. It can work as a binding agent. It can be used, at great effort, to make sulfuric acid. It is, on occasion, used as a fertilizer, applied to soils that lack sulfur. Phosphogypsum is gypsum with heavy metals and radioactive elements attached to it. It contains the radium that killed Marie Curie, the uranium that built the green glass sea. It degrades over periods of thousands of years. It is virtually unusable. In the United States, giant mountains of phosphogypsum are piled hundreds of feet high, surrounded by dirt, and topped with ponds containing acidic wastewater, left over from processing. Each pond holds millions, sometimes billions, of gallons of water. Each stack encompasses hundreds, sometimes thousands, of acres of land. The creation of one ton of phosphoric acid leads to the formation of five tons of phosphogypsum. The phosphate industry in Florida has left behind a billion tons of phosphogypsum. It adds about 30 million tons each year, creating mountains that rise with rapidity and tower over the landscape where they sit.

At the time that Progress Village was constructed, a fertilizer factory and phosphogypsum stack were in operation nearby. Compared to present standards, the stack was relatively small, and in the 1970s, the fertilizer company Gardinier bought the plant and proposed the construction of a second, larger stack, which would reach a height of one hundred feet across an area of 326 acres. When it announced in 1981 that the new phosphogypsum stack would be situated directly across the street from Progress Village—and immediately next to an elementary school—the community responded with fury. Working with the advocacy group ManaSota-88 and the American Civil Liberties Union of Florida, residents of Progress Village resisted the approval of the stack. They were worried about the health effects associated with phosphogypsum, and they asked the oncologist Gary Lyman to study the risk. Lyman, a physician-researcher who worked at the University of South Florida and the Moffitt Cancer Center, studied the contents of the proposed gypsum stack and the chances they would make people sick. He wrote a report that detailed his conclusion. "One of the concerns I raised was if this was to get into the ground of the school, children being children are very likely to be exposed to it, either playing in the play yard or getting it on their hands and then into their mouths or their eyes. And so the potential for exposure of children in that area was probably of even greater concern than even the exposure to adults. And then again, if it was in the air, being inhaled, these particles get into the lung, and then can expose the lung tissue to ionizing radiation that's released. Or if it's consumed through the GI tract, radium is a bone seeker. So if material that contains radium gets ingested by water or food or just dirt, the radium will often find its way to the bone—and the bone marrow, of course, is where leukemias develop," Lyman told me. "The bottom line from my standpoint, both as a resident of the county and a cancer epidemiologist, was it would be better if they found a more distant site for a waste stack, and not in proximity to a residential, perhaps disenfranchised, community, and a school that children would be studying in and playing in if exposure occurred."

Lyman testified before the Hillsborough Board of County Commissioners. He gave the board a copy of his report. "They were attentive. I don't think there were any dismissive comments. But it was basically, obviously, just something they were taking and filing with whatever other information and testimony they were seeking. I gave my testimony and left," Lyman said. On August 20, the commission approved the stack. A year later, Gardinier filed for bankruptcy and Cargill acquired the plant. In 1993, Hillsborough County permitted the phosphogypsum stack to grow from 100 to 200 feet in height, and in 2000, the maximum was increased to 250 feet. "Anyone with a brain would know why they decided to put that gypsum stack there," Alfred Sheffield told me. "They're looking at a community that is very poor, that doesn't have a voice, and that doesn't have a lot of political clout. That's how these companies operate."

After studying the proposed gypsum stack at Progress Village, Gary Lyman went on to look into the cancer risks associated with phosphate mining across the region. In a 1985 study published in *The Journal of the American Medical Society,* he showed that high radium levels in well water in Florida correlated with elevated leukemia rates, which had increased by up to three times. This was a key finding because leukemia, unlike many other cancers, is rare, fast growing, and environmentally activated. It can serve as an indicator of localized cancer risk. In 1988, he joined a study, led by the USF epidemiologist Heather Stockwell, that found increased risk of lung cancer among people—particularly women—living in the Central Florida phosphate mining region. In a vastly understudied geographic area, both the 1985 and the 1988 studies were fraught with limitations: neither proved that phosphate mining causes cancer, though both indicated that living in the mining region does seem to correlate with increased cancer risk. "It can never be used to prove cause and effect," Lyman, who now works as a professor at the Fred Hutchinson Cancer Center in Seattle, told me, "but it did raise concern."

Levels of gamma radiation are significantly heightened in and around land that has been mined for phosphate, sometimes reaching more than sixteen times background levels, according to surveys made by the U.S. Department of Energy and its partner, Argonne National Laboratory. According to the National Academies of Sciences, even low levels of radiation can cause cancer: there is no clear threshold at which risk begins. Throughout the 1980s and 1990s, the U.S. Environmental Protection Agency conducted regular radiological testing of the Florida phosphate district. In a 2003 internal report, the U.S. Environmental Protection Agency concluded that people residing on reclaimed phosphate mining land—including forty thousand residences in Polk County—were at risk of radiation poisoning from radium.

Despite the presence of radiation, only limited study of the health effects of mining in Florida has been done. Generally, research funded by the industry indicates little risk, while outside studies draw connections between phosphate mining and poor health. "In the United States, counties with phosphate mining or processing facilities frequently also have significantly elevated rates of mortality from lung cancer," Robert Fleischer, a nuclear researcher at the General Electric Research and Development Center, wrote in a paper published by *Environment International* in 1982. Fleischer analyzed national lung cancer statistics and concluded, "The data presented raise the possibility that phosphate mining and processing lead to an increased lung-cancer mortality rate." A study of 3,199 workers employed in a Florida phosphate fertilizer factory between 1951 and 1976, conducted by researchers from the CDC and the University of Illinois Chicago, found increased lung cancer and leukemia mortality among industry workers when surveyed in 2011.

In 1975, the EPA announced preliminary results of a study indicating that, due to radon exposure, residents living on land mined for phosphate found their risk of lung cancer doubled. The Polk County Health Department, concerned about these studies, placed radiation monitors in 750 homes across the region. In 1976, then-president Gerald Ford took a campaign tour of Florida and learned that phosphate mining posed dangers to human health. That May, Ford called for a new environmental study, and the EPA halted further phosphate mine permitting in Florida. For two years, Ford's concerns were investigated. The resulting report, released in 1978, described four potential pathways for radiation exposure in general members of the community: air contamination by radionuclides in dust, contamination of groundwater by seepage of process water from slime ponds and chemical plants, radon contamination on recovered lands, and consumption of crop foods or cattle raised on reclaimed land. There was, however, no proof of imminent harm: "No activity of the phosphate industry has been proved to cause a radiation dose to the general population in excess" of the guidelines set for civilians. Following the EPA's study, mine permitting began again. The industry moved on. The third and final mine built around the Killebrews' house was approved that year.

In Progress Village, the evidence of elevated health problems is mostly anecdotal. No formal academic study of the community has been completed since Lyman's work in the 1980s. "It's continued to be quite understudied and probably deserving of more study. This is the problem. Uranium in phosphate ore, although at low level, is there forever, essentially. It's a problem that's never going away," Lyman told me. Beyond these scattered studies, anecdote does not provide a basis for proof. But the people who live in Progress Village invariably know people—many people—who are sick. Walter Smith, an environmental engineer with roots in Progress Village, said, "You might hear people talking about asthma, or they're having breathing issues, some sort of respiratory issues. But mostly what you'll hear is people talking about how the material from the stacks goes into the community and settles on the lawn furniture, or the cars, and they have to go out there and wipe their stuff down every so often. If that's what's blowing into the community and getting on their cars, it has to be what they're breathing." He also said, "Most people who live there know. They have an idea. But the issue is, what are they going to do about it? Most people can't just up and move. No one's been compensated enough that they can do that. So that is another issue that has to be addressed. How do we deal with this? How are we compensated for the damage that's been done?"

Norma Killebrew speaks with a drawl and carries a long wooden walking stick with her wherever she goes. In eighty years of life, she has caught mullet, taught Shakespeare, read Plato, and dressed many a wild boar. Her eyes are mahogany, her hair is white, and the sides of her glasses are colored like raspberries. The day we met, her lipstick matched her frames. By 2004, when IMC and Cargill had merged to become Mosaic, digging across the road from their ranch had begun. The land dried up and began to fly away. "It turned the soil into white sand. I have a picture of it, of a barbed wire fence three strands deep where the white sand's coming over the top. You know how small sand is, and you know what a barbed wire fence looks like. You know that's quite a feat. It looked like a desert. In fact, I talked to the county about it, couldn't get their attention, so when I sent them a letter, I put the picture of the desert on the front, and the secretary at the local environmental agency—when I called her one day and she found out who I was, she said, 'Ms. Killebrew, I didn't know there was a desert in Hillsborough County.'"

To try to stop the mining, Norma Killebrew spoke at hearings, organized her neighbors, wrote letters, and took classes that taught her how to test well water for radiation and chemical contamination. She cited federal antidegradation rules, the Florida Administrative Code, and water permitting manuals from the Southwest Florida

Water Management District. She submitted complaints to Hillsborough County, to Mosaic, to the Florida Department of Environmental Protection, and to the SWFWMD, which is generally referred to as "Swiftmud." She became a one-woman expert on water permitting, well digging, radiological testing, and the industry at large.

In early 2012, she contacted the Environmental Protection Commission of Hillsborough County with new photos of sand and dust covering her property. On May 1, 2012, according to a sworn affidavit made in 2013 and shared with the U.S. Army Corps of Engineers, she "sent another email to Ms. Hallgren, E.P.C.H.C., telling her about the tremendous dust cloud from the mining and including a photograph of the cloud of particulate pollution. These new photographs included a picture of an alligator covered in phosphate mine dust as it crawled around attempting to find water. Mrs. Killebrew also reported that her family was having severe sinus problems and throat and ear pain caused by the long funnel cloud of dust heading towards Ruskin." On May 12 she was told by the EPC that Mosaic had dug a water-filled pit in an attempt to deal with the dust. "The Killebrews blame a vast expanse of stripped land, at least 2,000 acres, dotted with towering piles of sand Mosaic dug up and stacked as draglines probed for phosphate on the north side of SR 674, a few hundred yards from their property," the *Tampa Bay Times* reported on June 7. As part of the 2013 affidavit, Killebrew wrote, "Often we are held hostage in our home, unable to go outside to work on our ranch or even enjoy our rural property because the mining dust is so thick we can't breathe."

The Killebrews' connection to phosphate mining was complicated. As Norma fought against the mines, John, an ironworker, helped build the factories where the fertilizer was made. There are four superphosphate factories left in Central Florida; John Killebrew worked in all of them. He also worked at the Coronet Phosphate Company factory, a Polk County facility that has since closed. During his time at Coronet, the factory produced waste high in arsenic, cadmium, and chromium. It repeatedly violated local air and water quality rules enforced by local governments and the U.S. Environmental Protection Agency. During an inspection of the site in 2003, the Florida Department of Environmental Protection found waste stored in ditches and unlined ponds. Worried that local wells would be contaminated, the Florida Department of Health began a groundwater sampling program that found elevated levels of arsenic, boron, cadmium, sodium, thallium, and radium in water near the site. A 2014 administrative settlement between Coronet Industries—successor to the Coronet Phosphate Company—and the EPA described the consequences of what was found: "Arsenic can be carcinogenic if ingested in sufficient quantities over long periods of time and can cause dermal lesions when ingested in drinking water. Cadmium and chromium can be toxic to kidneys if ingested in sufficient quantities. Ingestion of sufficient quantities of fluoride can cause brittleness in teeth and can also negatively impact reproductive health." Further sampling revealed high levels of radioactive isotopes in the soil and water both on the site of the plant and in the surrounding communities. The federal guideline for the presence of radium in the environment is five picocuries per gram. In private residential wells offsite, concentrations reached thirty picocuries per gram. In unlined wastewater ponds on-site, the number reached as high as ninety-seven.

In 2003, two groups made up of residents and former workers sued Coronet, alleging material harm from contamination at the site. The following year, the plant shut down. The cases went unresolved: the residents of Plant City were never able to prove that their ailments were the result of Coronet's pollution. In 2014, the EPA assumed oversight of the Coronet facility, and in 2017 the Florida Department of Environmental Protection took control. By then, John Killebrew had been sick for decades. "He was forty-nine years old when he started having stomach issues. In 1990, he got sick. His stomach was hurting—he was in agony. I took him to the Brandon hospital. They told us they thought his colon had ruptured. They ran tests, and they found a mysterious clear fluid in his body cavity, but they couldn't figure out what was wrong with him. Two weeks before he got sick, he had started bleeding out of his nose every day he was at work at the Riverview plant. We had no idea what this meant. He went downhill a little at a time after that. We never found out what the fluid in his body cavity was. He told me his nose was bleeding every day. I begged him not to go back, and he didn't. He quit his job," Norma Killebrew said.

Over the course of the twentieth century, the mines of Florida assumed a prominent position among a global pantheon of phosphate producers. Processed into superphosphate, Florida's rock was shipped throughout the country and the world. By the turn of the millennium, the mines were producing as much phosphate every year as they had done in total when Roosevelt addressed Congress in 1938. In 2018, 25 percent of all phosphate rock produced in North America came from Four Corners alone. In 2019, wracked with cancer, struggling with Parkinson's, in pain and unable to breathe, John Killebrew died. "He kept saying he wanted to go," Norma Killebrew said, "but I wouldn't let him go. Finally, I said, 'If you want to go, you can.'"

In response to questions posed for this book, the Mosaic Company pointed to two local reports, written with its cooperation, that did not find a connection between phosphate mining and poor health outcomes in Florida. The company also referred to two independent studies that painted a more ambiguous picture. Although company representatives had readily answered questions related to other topics, they stopped responding when asked about the health effects of mining. In the years that followed, they left every email unreturned. I asked Walter Smith about the company's line. "We have yet to do the studies," Smith said. "The information that we have right now makes a connection. But in terms of actually making the connection from the standpoint of saying, 'Okay, these numbers here are higher than what you see in these other communities'—you hear things. People report it and say, 'Well, Miss Jean said that she has cancer.' And they turn around every so often. If you were to ask Alfred the same question, he'll tell you he turns around every so often and people are getting cancer. That is a red flag to me. If I were a government representative, I'd say, 'Get down there to Progress Village. I need you to get the health department. Get down there and start testing these people, and let's see whether or not there's some relationship.'"

Almost two centuries into the global project of phosphate extraction, Alfred Sheffield does not know if his neighborhood is dangerous or not. He does not know if the air is safe to breathe, the water is safe to drink, or the ground is safe for the children to play on. He goes to meetings and asks questions, but he does not receive satisfactory answers. Meanwhile, the phosphogypsum stack outside Progress Village is growing larger by the day.

THE PINEY POINT PHOSPHOGYPSUM STACK

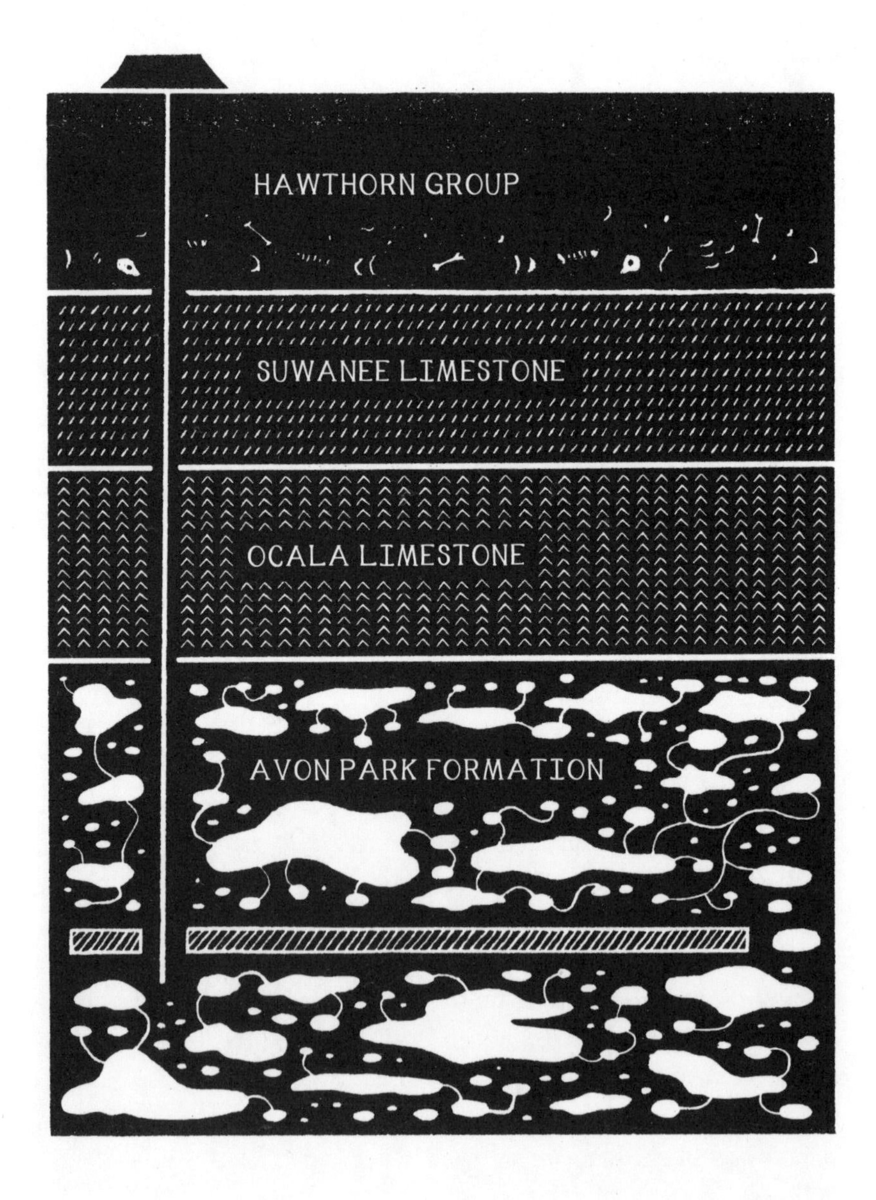

6

THE FLOOD

Before the national media came to this part of Florida, before the governor flew over in a rented helicopter, and before the public clamored with concerns about a "wall of radioactive water" descending down the coast of Tampa Bay, Jeff Barath knelt, alone, atop a man-made mountain, and he listened to the rhythms of the ground. It was three o'clock in the morning on the day before Easter, it was warm and humid and dark outside, and Barath was fighting to stop a crisis he long had feared would happen.

The mountain on which he stood was made of pieces of life that had been separated, processed, and rearticulated, their vitalizing parts removed and sold, their dangerous components left in place. It was a phosphogypsum stack, similar to those in Progress Village. It was piled up six stories high and possessed a series of indentions in the middle, at the top, in which were placed open-air ponds containing water. Both the contents of the waste and the contents of the water were remnants of phosphate processing, and both had come from phosphate rock. Here, on the edge of Tampa Bay, the toxins of millions of years, extracted over a period of decades, had been used to build a single hill. This place, plagued by shoddy construction, had frequently threatened to fail during its fifty-year existence. And Barath had been there for decades, laboring to keep the mountain in one piece.

That early Easter morning, in 2021, Piney Point was teetering toward disaster yet again. At three o'clock, Barath began walking the perimeter of its peak; it would take him five more hours to finish his circuit. He would depart, hoping to go home after several days of close monitoring. He knew there was a leak somewhere in the ponds atop the stack, and he had found the corner where he thought it might be. If the leak could be identified, the hole could be plugged before the problem got more serious. As he headed down the grassy slope, however, he saw water gushing out the side of the stack. He stopped and took a video. He would not be going home that night.

In the weeks that followed, the crisis would be known around the world. The stack was sixty feet high and made of mildly radioactive solids. The ponds were perched on top—put there out of convenience—and they held three-quarters of a billion gallons of water left over from phosphate processing. An implosion of the overflowing stack was possible, risking lives, homes, and the region's economy. Huge amounts of contaminated water would be siphoned into Tampa Bay in order to try to avert a widespread flood, and the ecosystem would suffer calamitous consequences from this disposal. Many of those consequences are still unknown today.

In the months that followed, there would be questions of how the stack came to exist in this particular location, questions of why its owners were not experts in the field. The people who owned the stack had profited enormously off of their investment while the risk of its collapse had only grown. There would be questions of why the government had allowed this to happen. There would also be, in the midst of all the chaos, a salient fact: this was only one such site. Across the state of Florida, there were twenty-four others, including the one at Progress Village. Across the United States, there were seventy. At each of those sites, a catastrophe could happen. A sinkhole, an earthquake, a hurricane, or simple human error could doom each facility, risking lives nearby. In many cases, the risk would be far greater than at Piney Point, which is considered small, as phosphogypsum stacks go: newer ones are often ten times taller and hold billions of gallons of water each. Some of them have failed before, dumping millions of gallons without oversight and affecting the health of those nearby. And even stacks that have not failed still pose a continued risk. They will be around forever, and they must be monitored and maintained. When the companies that own them abandon responsibility, taxpayers fund the upkeep on the site. In 2021, when Piney Point failed, the gravity of this situation became clear. ("Piney Point is what happens when you let the sharpies and hustlers have free rein. Money talks and the environment walks," the author Stephen King would write.) Even now that the crisis has ended, Piney Point—and six dozen other places like it—continue to exist, as they will for longer than anyone will ever live to see.

Phosphogypsum stacks have, over time, proven difficult to contain. The fears of Alfred Sheffield have been validated. The immense weight of each stack bears down on the cavernous karst topography that makes up Florida's base. Walls shift. Sinkholes form. The limestone cracks under the added weight. Water, held atop the surface, gushes down into the earth. In 1973, when the EPA investigated the Florida phosphate industry, it noted:

> Reaction between the acidic gypsum transport water and the limestone aquifer, together with solution cavities in the aquifer, has resulted in the movement of highly radioactive water into the shallow ground water of a large portion of Polk Country [*sic*]. . . . Many plants, especially those in Florida, are located on areas that are underlain by limestone. The acidic gypsum pond water can react with limestone resulting in the development or enlargement of cracks, caverns, and other solution features. This, in turn, permits the movement of contaminated water into the ground water. Many plants which report a negative water balance on the contaminated-water system are actually discharging radioactive waste waters to aquifers.

In 1994, a sinkhole was discovered beneath a phosphogypsum stack outside Mulberry, in Polk County, at a phosphate plant that was then owned by IMC-Agrico. The phosphogypsum in the stack was piled 200 feet high. The sinkhole measured 160 feet in diameter and dropped 400 feet into the ground. Millions of gallons of water drained away.

The industry tried to improve its management, but the problems continued. In 1997, a phosphogypsum stack owned by Mulberry Phosphates released 56 million gallons of water into Florida's Alafia River, effectively destroying a forty-two-mile stretch of water from the stack to Tampa Bay. More than a million fish were estimated dead as a result. In 2004, the Riverview phosphogypsum stack—the one near Progress Village—spilled 65 million gallons of contaminated water into Hillsborough Bay when Hurricane Frances broke a hole in the side of the stack. Portions of the bay were badly damaged. In 2016, a sinkhole opened beneath a phosphogypsum stack near Mulberry, Polk County, that was located a mile and a quarter away from the site of the 1994 disaster. This hole, estimated at 100 feet in diameter, consumed a vertical column of phosphogypsum 190 feet high. The entire contents of the 78-acre pond—215 million gallons of process water—disappeared down the hole. The Mosaic Company, which has operated the plant since 2004, did not notify the public of the sinkhole. A television news crew broke the story of the leak nineteen days after company engineers had privately identified its presence. Mosaic apologized, and Governor Rick Scott pushed for a state law to make the company publicize future leaks. "It does not make sense that the public is not immediately notified when pollution incidents occur," Scott told reporters at the Mosaic facility. Today, Mosaic is required to notify the government within twenty-four hours of a future leak, and the government is required to notify the public twenty-four hours after that.

Following the 2016 sinkhole, residents living near the phosphogypsum stack noticed oddly colored water with strange odors coming out of wells, and they reported skin rashes and new illnesses, particularly among children. Three residents sued Mosaic in 2016 but withdrew their lawsuit a year later. Although the depth of the sinkhole was not known—it was at least three hundred feet deep—and the path of the toxins leaving the site could not be traced, Mosaic maintained that the phosphogypsum and process water did not affect local drinking water supplies. It took Mosaic two years and $84 million to plug the hole. Officially, the water did not pollute the aquifer: it only disappeared.

In 2015, the U.S. Environmental Protection Agency fined Mosaic $1.8 billion in response to its poor handling of 60 billion pounds of hazardous waste spread between six phosphogypsum stacks in Florida and two in Louisiana. It was the largest environmental consent decree in U.S. history to be filed under the Resource Conservation and Recovery Act, a 1976 law that governs disposal of hazardous wastes. It stemmed from the revelation, made by inspectors working for the EPA, that Mosaic was illegally mixing chemicals at its phosphogypsum stacks. Under the terms of the settlement, Mosaic agreed to invest $630 million to fund future maintenance of its stacks.

In December 2018, a farmer in St. James Parish, Louisiana, noticed that a strip of land beneath his sugarcane—2,000 feet by 100 feet—was bulging. His farm was located next to Mosaic's Uncle Sam phosphogypsum stack. At Uncle Sam, phosphogypsum is piled 200 feet high and topped with a 960-acre lake of acidic process water. Mosaic's engineers soon discovered that the stack was shifting by more than half an inch daily, the result of heavy pressure exerted by the 700-million-gallon reservoir perched atop the stack. The company, fearing that the pressure of the stack could cause a collapse, began draining the reservoir and brought in dump trucks to pile dirt against the wall. The stack stayed still; its water was pumped into the Gulf of Mexico.

At Piney Point, the problems started early on. The facility was built in 1966 on a thousand-acre piece of land near the mouth of Tampa Bay. It was meant to process phosphate rock into phosphate fertilizer, which would be loaded onto ships and sent around the country and the world. Here, for thirty years, fertilizer would be made essentially the same way it had been made by Lawes, by Packard, by Fison, Mapes, and De Burg. Phosphate rock would be mixed with sulfuric acid, which would separate the phosphorus from the other pieces of the rock. Fertilizer, and wastes, would be the result of that process.

The history of Piney Point epitomizes the relationship, developed in the United States during the twentieth century, between finance and agriculture. The story of this facility is not one of chemistry or biology. It is a financial story—a tale of debt and bankruptcy, of investment and foreclosure, and, above all else, of profit. The factory had been built by the Borden Chemical Company, better known for producing milk. In 1970, Borden was caught and fined for dumping wastewater into nearby Bishop Harbor, causing fish kills. In 1989, residents were evacuated when 23,000 gallons of sulfur leaked from the site. In 1991, a cloud of sulfur gas killed three employees at the plant and sickened at least thirty others. The FBI investigated the sulfur leaks, seizing company records in a 1992 search. By the time Piney Point closed, its ownership had changed multiple times, culminating in a bankruptcy that left the site in the hands of a shell company called Mulberry Corporation, which was run by out-of-state investors. The plant closed briefly, then reopened. In 1999, it finally shut for good.

When production at a superphosphate plant ends—when a phosphogypsum stack is no longer growing with additions of new waste—the company that operates the plant is required to oversee the safe closure of the stack, which includes a multi-year process of removing contaminated water and a multi-decade responsibility to monitor the phosphogypsum for possible leaks. At the beginning of 2001, less than two years after production ended at Piney Point, Mulberry abandoned the site. With two days' notice, the U.S. Environmental Protection Agency took control of Piney Point. EPA engineers found a facility that struggled in nearly every respect. It leaked. Its water was untested. There was no plan for closing the site, which contained 1.3 billion gallons of water perched atop hundreds of acres of mildly radioactive phosphogypsum stacked twenty feet high. The lakes atop the stacks were twenty-seven feet deep, and they were constantly at risk of overtopping the dikes that held them in. They were open to the air, which meant that their water levels were constantly rising due to rain. A breach would be catastrophic: if water went over the top, it would quickly erode the dirt and flood nearby land with acidic, nutrient-rich, mildly radioactive water that could hurt people, destroy property, and cripple Tampa Bay. After Mulberry left Florida, the public assumed full responsibility for the site.

In 2003, two years after Mulberry declared bankruptcy, the Florida Department of Environmental Protection brought in Jeff Barath, an engineer, to look for solutions to the site's water problems. At Piney Point, Barath faced a Herculean task. The first two contractors to oversee the site had each declared bankruptcy. The site was poorly constructed and ill maintained, and its environment was not well suited for it. Ideally, the ponds would evaporate over time, but in Florida's humid climate, evaporation happens slowly. The occurrence of rainstorms and hurricanes meant that the balance of water in the ponds went up rather than down. Nine months after the government took over the site, rain from Hurricane Gabrielle had packed the ponds with too much water, forcing a release of 10 million gallons into Bishop Harbor. Soon afterward, FDEP had obtained a permit to release wastewater into the Gulf of Mexico. A barge carried loads of treated wastewater forty miles offshore, 7.5 million gallons at a time. Over a period of five months, 248 million gallons of polluted water were dispersed throughout the Gulf. Nevertheless, in 2002, following a thirteen-inch rainstorm, Piney Point's ponds again nearly overflowed. The ponds, which were supposed to be empty, were instead as full as ever.

Two years after Barath was hired, the State of Florida committed $140 million to a long-term project dedicated to fixing and closing the site. Two ponds were closed, 440 acres of plastic liners were installed, and a system was built to keep the water that dripped through the phosphogypsum from seeping into the ground. Between 2001 and 2007, the FDEP, with the guidance of Jeff Barath, removed 2.35 billion gallons of water from the ponds. (As a part of that project, 1.1 billion gallons of nutrient-laden water from Piney Point was released into Tampa Bay, causing nutrient pollution and algal blooms. In early 2004, FDEP stopped the practice.) Rain, however, continued to fall, and the ponds atop the stacks began to fill again. "We started out with sixty-four million dollars, we spent one hundred ninety-one million dollars, and we never finished," Barath told me later.

In 2006, for a price of $4.3 million, the Piney Point site was purchased by HRK Holdings, a company formed by three men who were new to the phosphate industry. William Harley III, the lead investor in the group—the H in HRK—was a prominent Denny's investor who owned four Hooters restaurants in New York and controlled a 46 percent stake in Frederick's of Hollywood, a lingerie retailer the *Los Angeles Times* once wrote sold "women's clothes from a man's point of view." HRK hoped to profit off Piney Point, which was located on 676 acres of valuable land close to the port of Manatee. If maintenance costs could be controlled, then a profit could be made by leasing some of the land and, eventually, closing the site. As part of the transaction, HRK agreed to set aside $15 million for maintenance.

After acquiring Piney Point, HRK struck a deal with the port of Manatee to place dredged material from Tampa Bay into one of the ponds above the gypsum stacks. The port of Manatee paid HRK $3 million for disposing of the waste. In 2011, when the dredged material was added to the stacks, the new liner broke immediately. Through a hole in the liner, 35,000 gallons of water rushed every minute. To prevent a total failure at the stack, Barath, who had recently been made manager of the site, redirected 170 million gallons of water into Tampa Bay. "I was terribly afraid, because you have thirteen hundred single family homes that are at lower elevation to the south of this facility. And rather than have a wave of water going that direction, I wanted to try to control it on-site. When this took place, this is what really put everything into chaos and craziness."

The 2011 leak came at a trying time for HRK. In 2005, Harley had been sued by the Claude Worthington Benedum Foundation, a Pittsburgh-based community development organization, which alleged that Harley had refused to return a $2 million investment made by the foundation in Harley's hedge fund, Fursa Alternative

Strategies, in 2004. As the lawsuit wound its way through the court system, it left a paper trail of troubles at HRK and Harley's other assets. Between 2008 and 2012, according to court filings, the value of Fursa's assets declined from $306 million to $125 million. In 2011, facing repeated requests from the foundation to withdraw its money from the hedge fund, Harley disappeared. He had no business address and, according to court documents, was operating his hedge fund out of the basement of a Hooters restaurant. As HRK struggled, Gary Kania departed the company and Scott Rosenzweig died. Only the H was left.

According to a subsequent investigation by the *Sarasota Herald-Tribune,* money from the Benedum Foundation was used to fund operations at Piney Point. Foundation officials speculated that their investment may have been used to help purchase Piney Point in the first place. By 2012, the foundation was told that the value of its investment had declined by 96 percent. By then, HRK had borrowed $17.5 million from AmSouth Bank, listing the Piney Point facility as collateral. "HRK, being a hedge fund group, had already made all their money back on the facility," Barath told me. "By the time I was done tearing down all the former plants and everything else here, it was appraised at over fifty million dollars. And they were able to pull twenty-plus million out of it that I don't know where that money went. You'll have to talk to the folks up in New York. At that point, the site was grossly underfunded."

In 2012, HRK declared bankruptcy, just as other owners of the site had done before it. Piney Point went into foreclosure, and parcels of land were sold to the highest bidders. The industrial facilities were acquired by other businesses, but no one bought the phosphogypsum stack itself. HRK retained ownership of the stack, and Barath remained in charge. Despite the work that had been done, the stack had not been closed—and the water at its top was still accumulating.

If, by some matter of happenstance, you were to meet Jeff Barath, you would probably like the man. He listens to people. He uses the words *ma'am* and *sir.* He praises his colleagues and employees. He operates with an earnestness that is underlain by years of frustration and fatigue. Even after months of death threats and abuse—his house was vandalized, his phone was doxed, and his family was threatened with violence—he remains quite possibly too polite. We met because I trespassed at a Piney Point construction site; he apologized to me. When we spoke a second time, he was seated in the aging trailer that contains his office, leaning forward, his elbows propped on a desk, his hands clasped before his face. He has dark, penetrating eyes, black hair, and a goatee. He wears glasses, hoodies, and oxford button-downs. At the end of 2020, he began to notice changes in the stacks at Piney Point. The chemical makeup of the water seemed different. The stacks had long been full. "We are quickly running out of process water capacity inside of the Piney Point facility," he had told the Manatee Board of County Commissioners in September. "We are incrementally going the wrong way." Four months later, in February 2021, he told the board, "Based on our current water volumes, action is needed. Now."

The problem was the same as it had always been: the ponds atop the stacks were open to the elements, and they filled with rain faster than they could be emptied by other measures. If the ponds grew larger than their capacity, they would overflow, potentially devastating the stacks. Even without overflowing, however, they exerted enormous pressure on the phosphogypsum stacked beneath them, creating the potential for leaks. Ten years after the 2011 leak, Barath worried that something similar might happen again. Working on a limited budget, he searched for ways to prevent catastrophe, but he lacked the resources to lower the water levels. "You see me in those county commission meetings about every three or four months, briefing the county commission about what was going to happen if we didn't get any help. Manatee County said, 'This isn't our problem. It's somebody else's.' And I already knew the people I worked for were Wall Street sharks," Barath said. "That's what they are. And I will tell you they were scumbags. Absolute scumbags."

In the middle of March, Barath began to notice unusual water chemistry readings in a part of the largest pond at Piney Point. Rapidly changing acidity in a wastewater pond indicates the movement of water through the pond, likely because of a leak. Barath worried that water flowing from the pond might collect in the phosphogypsum stacked below it, causing a watery bulge and perhaps an implosion, which would drain the pond and flood the area around it. Fearing this result, Barath began drilling holes through the liner below the pond to relieve the pressure on the gypsum stacks. Then he set about determining the location of the leak.

The plastic liners that cover the bottoms of the ponds extend up the embankments that surround the water. They are black, and in the sunlight, as they heat, they loosen and move, prompting slight vibrations. As night falls, they tauten and cool. In darkness, they reflect the movement of the water. A leak, consisting as it does of water moving quickly through a hole, would cause the liner to vibrate. On Easter weekend, Barath purchased a stethoscope from a drugstore and waited until the dead of night. From three until eight o'clock in the morning, he walked around the perimeter of the pond, listening for strange rhythms in the liner. In one corner of the pond, he heard a different rhythm coming from below—possibly the leak. That morning, he found water gushing out the side of the phosphogypsum stack, far below the level of the surface of the pond. There were rapids descending the stack. "That was scary to the rest of the world," Barath told me, "but that really made me feel good, because the system was relieving itself exactly as it was supposed to."

The leak relieved the pressure on the stack, and the water would be kept in an emergency pond on-site—but the leak was growing uncontrollably, and the risk of collapse became real. Barath needed to get more water out immediately. That Saturday—the morning of April 3—he requested an emergency final order from the Florida Department of Environmental Protection to allow 480 million gallons of water to be pumped into Tampa Bay. On Sunday, the order was granted. Barath's team began to set up pumps and pipes to divert water into trenches that led to the bay. Three minutes before four o'clock on Tuesday afternoon, the first water began to siphon out of the stack; it reached Manatee Harbor nineteen minutes later. Barath was pumping water at a rate of eleven thousand gallons a minute, and a rush was on to empty the stacks before they broke. The water in the NGS-South pond was twenty-seven feet deep. It was dropping by a foot a day.

That weekend, Governor Ron DeSantis declared a state of emergency for Manatee and its neighbors, Hillsborough and Pinellas Counties. The side of the stack was shifting, and engineers worried that this indicated an imminent structural collapse. Residential areas near the site were evacuated. On Sunday morning, as the FDEP was granting Barath's emergency order, Scott Hopes, the acting administrator of Manatee County, was speaking to reporters: "We're down to about three hundred forty million gallons that could breach in totality in a period of minutes. The models for less than an hour are as high as a twenty-foot wall of water." The Manatee jail, which is located two thousand feet away from the phosphogypsum stack, was not evacuated, although houses near it were. According to the county, inmates had been moved to the second story of the building, and flooding at that location was not expected to exceed five feet. A headline in *The Guardian* read, "Florida Emergency as Phosphate Plant Pond Leak Threatens Radioactive Flood." In *The New York Times:* "Florida Officials Warn of 20-Foot 'Wall of Water' If Reservoir Breaches." Water was leaving the stack at a rate of 35 million gallons a day.

On the afternoon of April 5, amid concerns about a possible second leak, the evacuation area around the site was expanded, and inmates were taken from the jail to a safer site. Ron DeSantis arrived and posed for photographs looking down at the stacks from a rented helicopter. He stood for pictures with a pair of expensive pumps he had brought, though the pumps did not ultimately function. ("He wanted to make a big political statement," Barath said.) Engineers and officials from the Army Corps of Engineers and the EPA were on-site. Drones with thermal imaging capacity flew over the stacks, monitoring for additional leaks. On Thursday, the eighth of April, Barath returned to the Manatee Board of County Commissioners and spoke about the dangers to the stacks. "I have spent most of my days and nights constantly monitoring all aspects of this gypstack system and identifying failure points in it. I identify the failure points, and they happen constantly, hourly," he said. "We are working on this twenty-four hours a day. I was home yesterday for the first time in seven days, for five hours, to take a shower and change clothes." He spoke for almost an hour, cried three times, and was interrupted when a consulting engineer broke in to inform him that the pipe taking water from the reservoir had sprung a leak. From the lectern, Barath confirmed an order to shut off the siphon. Barath left the meeting, and Mike Kelley, an engineer consulting at the site, testified before the board. The attitude since Sunday, Kelley said, "was get the water off the stack now, immediately, to keep it standing." Kelley described potential consequences of a breach: flooding on streets, damage to U.S. Route 41, losses of property by homeowners to the south. In the path of the water would be 316 families, the county jail, and a natural gas plant that provided the county with electricity. Soon after the hearing, the siphon would be restored, and its capacity increased. "We were discharging twenty-four thousand gallons a minute to drop the elevations inside that Gypsum Stack South so that we could stabilize it and avoid a catastrophic failure," Barath told me.

"Wastewater from Piney Point has Tampa Bay on edge," CNN reported.

Barath and his team had brought in a submarine to locate the leak within the pond, and then they had designed a steel plate to cover the hole. The plate had an area of one hundred square feet and was an inch thick. Divers lowered the plate down over the hole using undersea balloons. They filled the hole with aggregate, cemented the plate with neoprene, and pumped seventeen thousand cubic yards of sand on top to act as ballast. They plugged the leak.

The stack did not collapse. By the end of April, discharges into Tampa Bay had stopped. "If you want my true opinion, there was no point during any of that operation, or during any of that event, when I thought the gypstack was going to fail," Barath told me. His team had pumped 215 million gallons of water into Tampa Bay. The water was nutrient-rich and somewhat acidic. It was not, according to the

FDEP, radioactive. The acidity may have killed manatees, poisoned fish, and disturbed seagrasses. The nutrients fed a bloom of algae in Tampa Bay. Biologists do not know the extent of the damage.

Recriminations flew in the aftermath of the leak. Environmental groups sued the State of Florida. The State of Florida sued HRK. Environmental groups sued HRK. HRK sued the State of Florida. *Sarasota Herald-Tribune* writer Chris Anderson called HRK's handling of Piney Point "one of the most disturbing swindles in the history of the state." "I'd love to string them up, but the reality of the situation is they have one shell company after another. These guys are rich and smart and they know what they're doing," Manatee County commissioner Kevin Van Ostenbridge told the *Bradenton Herald*. "It'll be like getting blood out of a turnip. We'll never get a nickel out of them." Prior to the publication of this book, HRK was unreachable for comment. Two emails sent to Greenrose Holdings, a new venture run by Harley, were not returned. (Recently, Greenrose Holdings filed for bankruptcy in New York.) Harley has not spoken publicly about Piney Point since the leak. In late August 2021, five months after the leak began, responsibility for the site was taken away from HRK and given to a court-appointed receiver, and Jeff Barath was left in charge of the site's management. Barath has more than one stack of lawsuits sitting on his desk in the trailer. He warned me, when we met, that our emails would likely end up subpoenaed. He does not have his own lawyer.

"I think that it is managed very responsibly," Barath told me, elbows propped and hands clasped, that day in his office at the end of 2021. "How it is managed now, things that we learned from this site, as far as the lining and the permit requirements and the expansion aspects and the geotechnical—I think that we manage this far more responsibly than anybody else on this planet." As a result of the decades-long catastrophe at Piney Point, Florida regulations governing phosphogypsum stacks have changed. There is now a system in place for closing a stack. Phosphate companies are required to pay money into a fund that would, theoretically, offset the cost of the closure.

Barath views Piney Point as a test case, a calamity whose misfortune has guided a reshaping of policy toward phosphogypsum stacks, leaving Florida safer and the phosphate industry more responsible in the long term. "I really want the real story out there. I am a poor scientist. I am a single dad. I have two teenage daughters," Barath said. "I am grossly underpaid, highly overworked. I work at minimum six days a week. But I believe in what I do. I also believe that I make a difference in my community and my environment, but also in my industry. And that is the only reason why I continue, because I want this done right." Six days after he told me that, the state would approve the closure of the property. Five days later, Piney Point's final journey would begin.

Following the leak, the Florida legislature authorized $171 million—some of them federal coronavirus relief funds—to finally close the phosphogypsum stack at Piney Point. The water would be emptied. The ponds would be closed. A team would stay for fifty years to monitor the chemistry of the groundwater, and then the phosphogypsum would be sealed and the stack would be left essentially forever, a looming mound in the flat landscape of Florida.

When I visited, eight months after the leak was plugged, Piney Point was quiet. Barath's team was girding for the rainy season. They were working on a plan to close the site. Half a billion gallons of water were left in Piney Point after the crisis days of 2021, and on December 16, 2021, the FDEP gave Manatee County a permit to build a well to pump that water underground. This move was a culmination of a decade of study by Barath. There were three possible ways to remove the water. The first was to evaporate the water by spraying it into the air. To test it out, Barath built evaporative systems on one pond and concluded that they could not function on a larger scale: in the humid Florida climate, not enough water could be removed. We passed by those systems as we drove around the site, fields of pumps spraying toxic water into the air. The mist, vaguely radioactive, glimmered in the sunlight. If Barath did not lower its volume on windy days, it could kill plants and sicken people. Even under careful management, he pointed out, the grass along the edge of the pond was dead.

The next option was similar to what had been done two decades earlier, when the government first began to drain the ponds at Piney Point: the water could be pumped into Tampa Bay. It would be treated for acidity and tested for radioactivity, but it would still contain the nutrients that would cause the bay to bloom with excess algae, damaging the ecosystem. "I did that from 2003 through 2007. I saw the impacts," Barath told me. He did not wish to do it again. The final option was the hole. Deep well injection—a pipe extending 5,500 feet downward that would pump all of the contaminated water into the lowest level of the aquifer, four aquifers down, where the water is brackish and, Barath insists, worse for drinking than the water in his largest pond. The Piney Point water would stay separated from the aquifers through which it passed on its way down, in part through the construction of vertical concrete barriers and an adjacent monitoring system. When Barath proposed his plan to the Manatee County Board of Commissioners, Priscilla Whisenant Trace, a commissioner, demurred. "We don't know fifty, sixty years from now what it's doing down there," she said at the commission's September 2020 meeting. The other commissioners shared her misgivings, as did advocacy groups and many members of the community. Water migrates. Hydrology is complicated, and no one knew for sure where the Piney Point water would go. It could end up in the Gulf of Mexico. It could end up in local wells. It could do exactly what its planners hope, and move slowly through the brackish water of the Lower Floridan aquifer for millennia, losing its human elements and becoming interchangeable with the water of the ground.

Following the leak in 2021, Barath convinced Manatee's leadership that deep well injection was the county's best worst option, and in December, the state gave Barath a permit to build the well. "Today's decision by the Florida Department of Environmental Protection to allow deep-well injection of Piney Point's hazardous waste is beyond reckless—it's complete negligence of their responsibility to defend our environment and citizens," the state's agriculture commissioner, Nikki Fried, said in response. On December 22, the boring started. A year and three months later, on April 3, 2023, the well began operation, draining Piney Point's wastewater at a rate of 1 million gallons a day. The water journeyed deep underground, past the Hawthorn Group, where phosphate rock once formed, past the Suwannee Limestone, which is located three hundred feet below the surface, past the Ocala Limestone, on which the Suwannee sits, and into the Avon Park, the Oldsmar, the Cedar Keys Formations. These are the layers at the heart of the aquifer. They are old sedimentary rocks, cavernous limestones composed of dark, rushing rivers and pitch-black subterranean lakes. The water went far deeper into the ground than the phosphate had been before it was extracted. Florida's phosphate sits 14 million years deep in the geological strata. The contaminated water from Piney Point—the detritus of Borden's phosphate—would be pumped into strata of the Cretaceous, which began 131 million years earlier. It would take three years, according to the plan, for all of the water to drain away.

There are now seventy phosphogypsum stacks in America, including twenty-five in Florida. The Florida stacks are spread out between five counties, but the overwhelming majority of them are located in Polk County. The phosphogypsum stack at Piney Point is sixty feet tall. The phosphogypsum stacks that the Mosaic Company is now building elsewhere eclipse five hundred feet. The water in those stacks is not just greater in volume—it is far more toxic than the water Barath manages. A break in the dikes could mean death by flooding or by poison. Piney Point, tiny as it is, has cost nearly half a billion dollars in taxpayer money, and it has resulted in the release of almost 3 billion gallons of wastewater into Tampa Bay—and now, deep into the aquifer. The other stacks are owned mainly by Mosaic, their maintenance funded by ongoing mining. "And what happens after the mines are gone?" Dave Woodhouse, a hydrogeologist who lives in Manatee County, asked me. "I think you know just as well as I do that the way out is to declare bankruptcy, whether you're phosphate or whatever—that way you can walk away." Andy Mele, the Peace Myakka Waterkeeper, said, "I think that if Mosaic was to go under, deliberately or otherwise, the people of Florida would find themselves confronting a landscape of gypstacks with a little bit of money there, ostensibly to do mitigation, but nowhere near enough. And it would be Piney Point times twenty-four." Mosaic could, in theory, leave Florida after the mineable deposits are emptied—it could declare bankruptcy and walk away. The Mulberry Corporation did it: in 2001, when Mulberry left Piney Point to the state, its owner kept the profits of the business but not the liabilities. The people of Florida picked up the tab. Mosaic denies that it would do the same.

It is difficult to convey the immensity of a phosphogypsum stack. It resembles a low volcano. From the ground, it looks like a mountain. From the top, it appears to be a sprawling lake. If you stood on one side of the main pond at Piney Point and waved at someone standing across the way, you would struggle to be seen. The roads around the ponds at Piney Point are dirt tracks. The path to the top is steep and rutted. At the end of 2021, there were three ponds still in operation. The largest pond, the one that leaked, was filling up again. It had ducks in it. Barath took me to the edge of the stack, sixty feet up in the air, and we surveyed the flat landscape that surrounds the facility. "Let's get up here to the top of the gypstack and get your bearings," Barath said. "At the southern property boundary, see the transmission lines? Those transmission power lines come out from the Parrish power plant. You can actually see the stacks off in the distance. It's kind of hard to see. They're over here, if you know what I'm looking for. My entire concept is that I'm going to turn the gypstack into a giant solar farm, because these are the transmission lines coming right out of the power plant. See the Australian pines? That's U.S. 41. Of course, you look directly across the street: that's the port of Manatee. See the Skyway Bridge over there? It's kind of a foggy day, so it's hard to see, but you have the city of St. Petersburg right over here, and Tampa over here. Around this entire gypstack, in all these drains, you have an under-drain system. And that is collecting any of the porewater that is draining out of the gypstack, and it is coming into a pump house over there. Then we are collecting it at about one hundred thirty-four gallons a minute right now. Putting that water into that pond, treating that water, then treating that water, then putting it up into the reservoirs up here. It's a heck of a neat project, but it is a super-complicated project, to say the least. But it is a really neat facility. And from an environmental standpoint, it is really unfortunate, the things that have transpired here."

7

PEAK AND VALLEY

We wander on this planet, looking around with trepidation, and we try to justify our place here. We lay out how it happened. There was an egg, a dream, a metamorphosis of worlds. The earth was only oceans; a diver came. A woman in the clouds fell groundward toward a group of water-dwelling animals: they dove into the sea, collecting bits of mud that then "grew into what is now North America." Elsewhere, "first ancestors . . . thrust down the jeweled spear of Heaven and stirred it about in the sea," and it reemerged laden, forming islands, creating Japan. When a creator "sent various animals to find the mud necessary for the creation of the world," only "the lowly leech was able to swallow some mud and spit it into the creator's hand." The creator made the continent. In the Altaic tradition, the architects of earth were, at one point, geese. One dove into the water on a mission "to find rocks and earth with which to build the world." The world was water everywhere, the creators lonely. Minerals were brought up from the sea, and modern life began.

Of course those stories hinged upon a rock. Ours does too. Four and a half billion years ago, the earth was a ball of fire that transformed into a world of stone. The lava hardened; the magma cooled. Giant igneous slabs grew up around the planet. Within those rocks, elements were dispersed. They were created and laid to rest by random acts of fire and heat. As the lava cooled, conditions changed. The chaos of the early earth gave way to something more structured. The inanimate world animated itself. The cause remains a mystery, but the requirements are clear. Life had to form somewhere rich with the underpinnings of biology: energy from the sun, water from the earth, proteins, carbon, and phosphorus. But phosphorus was rare, and it was spread around the planet evenly, held within the crystals of the stone. "You find other elements in living things, but phosphorus really occupies a unique linchpin role in a lot of organic structures," Jonathan Toner, an astrobiologist at the University of Washington, told me. "People want to think about, *How could you get phosphorus reacting to produce biomolecules at life's origin?* But the problem with getting phosphorus in biomolecules is that, one, there's very little of it around in the environment. And two, it's not terribly reactive. So you have these dual problems, which are generally known as the phosphate problem."

It is a question of the chicken or the egg. In the early days of everything—when earth was a ball of fire that transformed into a world of rock—which came first: concentrated phosphorus or life itself? The gathering of phosphate into higher concentrations is the work of biology: the formation of whale falls, the creation of fertile ground. On a lifeless early earth, there were no such collections—and then, somewhere, there were. In a lake, perhaps—or a vent, or a geyser, or a shallow sea—life coalesced. There was a spark of something. Rocks were brought out of the water, and a world of beings bloomed. (Toner's research describes certain lakes whose chemical balance would allow phosphorus to remain free, slowly accumulating and becoming available for life.) The distribution of phosphorus was never the same again. Here was a mechanism for collection instead of dispersal, for the concentration of matter that had long been geologically spread. Living organisms would roam the world collecting phosphate and excreting it. We are the descendants of those organisms, tracking down our ancestors for sustenance. Henslow walked at Bawdsey, and others followed him—trepidatious, justifying. They wandered through catacombs, around mummies, over battlefields. They chanced upon deposits of the past. But the journey of change would far outpace Henslow's small discovery. (Creating earth, "God made a ball out of the soil and then fell asleep.") To be alive is to search for phosphorus; we are going deep and scooping mud.

At the western end of the Mediterranean, in the part of the sea known as the Alboran, a mountain crowns a peninsula that points toward Europe. Although the peninsula is part of Morocco, the city of Melilla, its main metropolis, is Spanish: owned by Spain, governed by Spain, part of Spain since 1497, when Isabella and Ferdinand sent a convoy of ships to take the port and secure the land around it for the Crown. For as far as a cannon could fire from the port-side fortress, Melilla would be part of the Spanish Empire. Spain never gave it up, and today the country's African foothold—one of two—occupies 4.7 square miles and still conforms to the historical arc of a cannonball.

The border between Morocco and Melilla is marked by three tall fences topped with barbed wire, rigged with electricity, studded with lookout towers, and populated by agents armed with guns and leading dogs. Beside the wall on the Spanish side is a low, sand-colored structure: the Centro de Estancia Temporal de Inmigrantes. The CETI, it is called—Melilla's refugee detention facility. In 2017, at the CETI, I met Muhammed Ahl Abhayah, who might reasonably be termed a phosphate refugee.

Melilla is a contradictory place. Roads abut the sea, winding their way up to the top of the old fortress, still preserved. There is a public beach, Playa del Hipódromo, that is fronted by weather-beaten hotels. Golf Melilla, the green that runs along the border, was memorialized in a series of photos that depicted a dozen migrants sitting atop the border fence while polo shirt–wearing people played golf below. The migrants were trying to cross the fence to place their feet on what is, technically, Spanish soil. Climb, jump, touch the ground, and declare asylum on European land.

Muhammed was from Western Sahara, a territory controlled by Morocco in a violent occupation that has lasted fifty years. The occupation started in 1975, when 300,000 Moroccan citizens marched through Western Sahara in a mass claim of ownership over a nation that had, until that moment, belonged to Spain. The Moroccan interest had begun, roughly speaking, twelve years earlier, in 1963, when researchers turned up phosphate in the little desert town of Bou Craa, in Western Sahara's northern region. The Moroccan government had a history with phosphate: Office Chérifien des Phosphates, the government entity that then controlled all phosphate mining in Morocco (and still does to this day), was founded in 1920, eighteen years before Roosevelt addressed the phosphorus issue in Florida; its first mine opened the following year. At the time, Morocco was a colony of France, which had conquered the country eight years earlier. By the time Morocco resecured its independence in 1956, two OCP mines had opened. When the rock was found in Bou Craa, the Moroccan phosphate industry was in full swing. In the years that followed that discovery, Morocco's government pushed publicly for Spain to cede the territory; in 1975, the country took its southern neighbor by force.

Western Sahara is a desert place, sparsely populated, and its wealth comes from the ground. There are oil wells offshore and phosphate reserves on land. Although 600,000 people live in Western Sahara today, hundreds of thousands more have been forced to seek asylum elsewhere. Spain, following the Moroccan march of 1975, gave up its land—although a confidential agreement, later revealed, reserved for the Spanish government one-third ownership of the phosphate company that mined Bou Craa. The people of Western Sahara—the Sahrawis—rose up to rebel against colonialism. Morocco won the war that followed, cracked down on the Sahrawi people, and regained control of the land that had been lost. During the 1980s, Morocco built a wall to separate colonial territory from rebel-controlled lands. It is the longest contiguous modern border wall in the world. The Sahrawi side of the wall is mostly barren, while the Moroccan side includes Bou Craa and the coast. For almost five decades, Morocco has enforced a brutal police state in the territory, subjugating the Sahrawi people and taking their wealth as its own.

Muhammed Ahl Abhayah, phosphate refugee in Melilla, grew up on the occupied side of the wall, in Dakhla, a peninsula attached to the Atlantic coast. A Sahrawi, he was tall and slender and left to study sociology at Cadi Ayyad University in Marrakesh. After graduating, he returned to Dakhla. As a sociologist, he always tried to view his surroundings objectively; he loved his home, and he resented the repressive colonial government that controlled it. In early 2017, when he was twenty-four years old, Muhammed joined the Polisario Front, a revolutionary force that fights on behalf of Sahrawi independence from Morocco. He led a platoon of soldiers. ("I gave so much to my country in order to reach independence," he told me. "Either we are an independent country or a cemetery visited by the world.") He soon became a target of the police, suffering threats and beatings that made him fear for his life. He fled Western Sahara, traveling to Agadir, a city in central Morocco, and flew north. He made his way to Rif, a region in northern Morocco that borders Algeria, then traveled to the border with Melilla, evaded detection, and passed through the wall in 2017; by summertime he was living in detention at the CETI. We met outside the facility's front gate. He introduced himself to this book's illustrator and me. He was lanky and serious, and other people respected him. The first time we met, he showed us his university diploma. He told us there was abuse inside the camp. He said all he wanted was to leave. "My country, Western Sahara, is one of the richest countries in phosphate," he told me later. "Morocco is currently exploiting it outside the framework of legal legitimacy. . . . It is an occupier."

About half of all Sahrawi people live outside their nation's borders, many in Algeria. Some, like Muhammed, have snuck into Melilla: about 50 percent of the refugees living in Melilla when we visited were from Morocco and Western Sahara. The global agricultural system relies on the production of phosphate fertilizers from just a few places around the world, and countries have little choice in choosing the sources of their fertilizer. New Zealand, a country with a reputation for upholding human rights, has been, sometimes, Bou Craa's largest customer. Morocco's regime in Western Sahara exists in the face of local resistance and international outcry—but, as the historian Lino Camprubí noted in a 2015 article in *Technology and Culture,*

"While not a single country recognizes Morocco's sovereignty over Western Sahara, most count Western Sahara's phosphate reserves as part of Morocco's."

According to the United States Geological Survey, out of 78 billion tons of mineable phosphate rock left on earth, 55 billion are located within the jurisdiction of the current king of Morocco, Mohammed VI. All told, Mohammed controls about 70 percent of the phosphate reserves left on earth. Things could happen. A flood in Morocco could affect food availability for the whole world. Political instability could disrupt the fertilizer supply chain, leaving hundreds of millions of people without food. Morocco could act as a monopoly, raising prices or extracting political concessions in exchange for phosphate sales. Regional controversies could evolve into global issues of concern in which Morocco holds a great amount of leverage. "They could essentially hold the world hostage in terms of phosphate," Eric Hiatt, a geologist at the University of Wisconsin Oshkosh, told me in 2020. "If something happened in Morocco—if Islamists took it over and said, 'We're not going to sell phosphorus except at a really high price, or we're not going to sell it at all,' the price just doubles or triples overnight. Which makes all food more expensive. It's scary in terms of global population, and maintaining seven point six billion, headed for ten billion, people. It becomes very precarious. If some little thing goes wrong, a billion people will starve. We're living on the edge."

No one knows exactly how much phosphate rock remains available on earth. Estimates have varied over time. In 1944, when the researcher A. N. Gray wrote a comprehensive account of the industry, *Phosphates and Superphosphate,* a rosy outlook pervaded his reports. The global phosphate mining regime he described was spectacular. The United States, people thought, held the future of phosphate. There were mines in the United States in Florida, South Carolina, Tennessee, Montana, Idaho, and Wyoming, with others planned in Alabama, Arkansas, Kentucky, North Carolina, Pennsylvania, Utah, and Virginia. Throughout the Indian and Pacific Oceans, thirteen islands, many of them extraordinarily remote, hosted mines.

There was phosphate on the banks of the Nile, near the shores of the Red Sea, and spread throughout Tunisia, Algeria, and French Morocco. On the Kola Peninsula, high up in the Arctic Circle, the Soviet Union managed a phosphate mine of volcanic origin. To access the rock, miners dug a twenty-mile system of underground railways, culminating in a tunnel through Kukisvumchorr Mountain that was designed to take the ore to the coast, where it was shipped through waters freed of ice by the Gulf Stream. There were an estimated 2 billion tons of phosphate present at the Kola mines. Gray evinced little concern about the future of the industry.

Since the middle of the twentieth century, however, the prospects of future mining have grown more limited. As richer reserves are exhausted around the world, the overall phosphate content of mined rock is dropping at a rate of 1 percent per decade. As the phosphate content of phosphate rock, which generally hovers around 30 percent, declines, the cost of processing rises, and the waste created rises along with it. Applying calculations of petroleum reserves to phosphate rock, the researcher Dana Cordell suggested in 2009 that the world may soon reach "peak phosphorus"—the point at which production declines and fertilizer rises substantially in price. "While the timing of the production peak may be uncertain, the fertilizer industry recognizes that the quality of existing phosphate rock is declining, and cheap fertilizers will soon become a thing of the past," Cordell's team wrote.

It is, of course, hard to predict the future. Higher prices could fund mines in more demanding locations. New technologies could make more rock accessible. Many phosphate deposits are located beneath mountains or developed land; huge amounts are hidden in the ocean. Theoretically, engineers could figure out a way to dig that phosphate out, although the logistics of ocean mining are exceedingly difficult. There is currently no indication that phosphate could be obtained in such a way. (This difficulty extends across minerals. There is enough gold in the ocean, Rachel Carson noted in *The Sea Around Us,* "to make every person in the world a millionaire"—but clearly, despite the overwhelming incentive, no one has managed to find a way to get it out.)

It seems a natural conclusion to two hundred years of increasing reliance that phosphate, a fertilizer that spurred efficiency and led us all to monocultures, agribusiness, and top-down farming practices, will ultimately see its sources reduced to a single company run by a king. Phosphate rock has made its way into our farms and food, changing the way we structure our society and the way we live our lives. The arc of phosphorus bends groundward, however, and its peak lies close ahead. Phosphate deposits, those hidden treasures of past times, have always come to tire out eventually. (Whale falls, after all, disappear in decades.) And we have taken what we found. England's deposits emptied soon after Henslow discovered them. Florida's rock will be the next to go.

To estimate the future timescale of its mining, the Mosaic Company divides reserve holdings by annual output. In 2020, when the company mined 17.5 million tons of phosphate rock globally, it reported to the Securities and Exchange Commission that it had 661 million tons of mineable reserves left in its holdings. According to its 2020 numbers, the company would be out of phosphate before 2060. "There is a long-term decline in the U.S.'s position as a phosphate producer relative to other parts of the world," Andy Jung, Mosaic's chief economist, told me. The numbers, he cautioned, tend to wiggle—new reserves can always be added to the list—but in recent years, nothing new has been identified. By geological necessity, the United States is ceding its position on the global phosphate stage. "That trend is probably going to continue," Jung said. (He, like others, stopped communicating when I asked about the health effects of mining.)

Faced with limited reserves, Mosaic has long acted aggressively. In 2004, when the company was formed, its leadership eyed an expansion to the south in Florida. The land the companies owned in Manatee and Hardee Counties had long been either farmland or natural habitat. Cattle grazed there, oranges grew, birds nested, and streams curved across the landscape. Quiet and rural, the land fit with its surroundings. Mosaic encountered opposition wherever it went, and the company responded with force. In 2008, when Hardee County's board of commissioners announced its opposition to a proposed mine, Mosaic offered to give the county $40 million, and the zoning request was approved. Later that year, when Manatee County's commissioners denied a new zoning change, Mosaic announced a $600-million lawsuit against the county, and the commissioners changed their minds. "Mosaic definitely comes into the room to expect an approval," Joe McClash, a longtime Manatee County commissioner who voted against the rezoning, told me.

In Hardee County, one in eight acres has been mined since 2008. In 2010, Frank Kirkland, a Hardee resident, was driving down the road near his house when he noticed draglines digging rock in violation of a court order. He pulled over and began to take a video. His son Stanley was with him. Two private security guards employed by Mosaic ordered Kirkland to stop filming. He refused. The guards' role was complicated: they were off-duty cops. In their dual capacities as Mosaic guards and government officers, the two men tasered Kirkland. "Responding to his father's pain, Stanley Kirkland then made demands to set his father free. In response to Stanley Kirkland's demands to set his father free so they could all leave, Defendant Wright came up behind Stanley Kirkland, grabbed his left arm, forced the arm behind his back, and placed Stanley Kirkland under arrest," the Kirklands' lawyer wrote in a lawsuit filed against Mosaic. The Kirklands complained of physical harm and intimidation and of the use of government authority for private ends. Their suit was unsuccessful.

In 2014, the Mosaic Company filed an application to rezone a fourteen-thousand-acre tract of land it owned in DeSoto County to allow for mining. The mine would be built on land straddling Horse Creek, which emptied into the Peace and was the river's last major tributary undisturbed by mining. In the land around the creek, under Mosaic's plan, phosphate would be mined for decades and would leave behind eight clay settling areas (CSAs)—the same kind of waste facility that caused the floods that had devastated the Peace River repeatedly during the twentieth century. Some of those facilities would be more than a mile wide, each one holding sludge left over from the initial processing of the mined rock. Horse Creek plays an important role in supplying drinking water for more than a million people. "Horse Creek fans out to the northward. It's got a million different branches and forks and stuff in its headwaters, and they all come together. It's at the point where they do come together where these CSAs are going to be located. At its best, at its mightiest—unless it's in full flood, in which case it can get a little scary—Horse Creek is thirty feet wide. It's a crick. It's a creek. It's not a river. And having these monstrous—these things are going to be sixty feet high, thirty feet below grade, for a total of ninety-some feet in depth. And they're going to be two-thirds to a full mile to a side, these big rectilinear clay settling areas," Andy Mele, the Peace Myakka Waterkeeper, told me. "As for the Horse Creek, it's done, it's toast if they build this stuff. The Horse Creek won't be anything you can refer to as a creek. It'll just be a thin industrial sewer."

In 2001, according to an industry-sponsored study, the division of phosphate rock from quartz and other minerals required about 120 million pounds of fatty acids and 240 million pounds of fuel oil annually. Roughly one-third of that oily mixture, the study acknowledged, "is unaccounted for and may be released to the environment." According to the study, these oils, acids, and salts released into the ground and local waterways each year quickly—one might say magically—biodegrade, presenting little cause for concern. The remainder of that mixture is held inside clay settling areas, where it remains the company's responsibility forever.

The DeSoto mine would be Mosaic's final one in Florida, and the company's request to rezone the land was quickly approved by federal and state authorities. The rezoning needed to be approved by DeSoto County itself, and a hearing before the board of commissioners was set for 2018. The community was divided about the plan. People were, understandably, scared. Brooks Armstrong runs a small farm on acreage he owns in Hardee. "All you have to do is go up there and see what's happened," he told me. "It's kind of a boom-and-bust thing. Once the mine has gone through there, the soils are depleted. You don't see a lot of real healthy agriculture. I go out there—I don't see cows. I see a lot of what looks to me like wasteland. . . . And it's currently in a boom period. When the mining is all done, the money will be gone. The money that they do benefit from goes into a few pockets and doesn't flow to the entire citizenry of the entire county." Molly Bowen lives in Arcadia, DeSoto's county seat. "We survived perfectly fine without Mosaic money for all those years. Why now?" she said. "They hype up the economic aspect of it, but that's not how it turns out. They end up ruining your town."

There is a history in Florida of companies mining phosphate, declaring bankruptcy, and leaving their waste in pits for the public to clean. I asked Andy Jung what makes his company different from those. "The short answer is that Mosaic is just a very transparent and ethical company and has endeavored to be so for quite some time. The differences between some of those other companies in the past and Mosaic are that we're operated with a much different mentality. There is more regulatory oversight, and that probably allowed companies in the past to skirt their responsibilities," he told me. "Mosaic tends to do what we say and say what we do. Be transparent and then execute."

The hearing in DeSoto began with a prayer. "Tensions are high today—that's no secret," a pastor told the room on the afternoon of July 24, 2018. "But, Lord, I pray that we would treat one another with grace, with love and care. That we would speak to one another and treat each other the way that you call us to be. And, Lord, I pray that all things in this room, that decisions are made that would first honor you and benefit the citizens of DeSoto County."

Over two days, five members of the public spoke in support of the mine, and seventy-five spoke against it. Of the five who supported the rezoning request, three were employees of Mosaic. Of the two supporters who were not employed by Mosaic, one said, "Man, it's just perfect where we live. And it is going to literally suck to have the mine there. But we've gotta have it." The other said, "If you deny their rezoning, I think you're trampling over the rights of all of us who own property and have been here a long, long time." Otherwise, the public did not support the mine. "This land is our retirement. When we're too old to work this land, it will finance our final years. If our property value drops because of mining, we are lost," a local farmer said. "I understand your sole requirement is to consider facts and numbers when making your decision, so I ask you: How do you quantify beauty? How do you put a number on happiness? How can you plot peaceful living on a chart? In what column would you place the satisfaction that comes from thirty healthy new calves or a successful honey harvest?"

A ship's captain named Paul DeGaeta was the seventh speaker to come to the front of the room on the first day of public comment. His biography had prepared him for this moment. He was four years old in 1958 when his family moved to Florida and built a house along the Peace. He remembers the feeling of life all around him at the river, and a sense that he was entering a brand-new world. "The very first time I peered over our dock," he told me later, "I have this magical vision of looking down in about three feet of water. It was high tide. From the dock, the water was clear. It had that brown tannic tint to it. But I could see the bottom, and the seagrasses were just moving with the tide. There were what looked like a zillion baby shrimp in there, bouncing around, and the barnacles on the seawall itself. I saw a school of mullet go by. I saw a stingray. I don't know how many times I've seen one swimming like that. It swam. It wasn't on the bottom. Probably because it was going over the grass. But I remember that. It's still seared into my soul. That's when the river entered my being."

When DeGaeta was eight years old, his dad gave him his first boat: an eight-foot Kury pram with a small outboard motor that could power the vessel up and down their little stretch of the Peace. He hunted fossils and arrowheads in coves along the river and caught mullet, snook, and blue crabs that his family sometimes ate for dinner. The Peace River is renowned for its tarpon—big, buoyant silvery fish—and his dad spent two years directing the International Tarpon Tournament. "There were always fish. You cast out there—you spent thirty minutes if you were going to catch something. And just to see that, see the tarpon rolling out in the river, it was just incredible," DeGaeta said.

On March 14, 1967, when DeGaeta was thirteen years old, he woke up to find that the river outside his house was covered with white dots and dead fish. The following day, the entire river turned a milky white. "It was just awful," DeGaeta said. "I loved this river so much." His family soon found out what had happened: a phosphate spill at a Polk County mine fifty miles upriver. A dam had burst at a clay settling pond, dumping a mine's worth of toxic sludge into the river. Upstream, the material had flooded the area completely, temporarily doubling the river's size. "There was phosphate sludge ten feet deep in parts of the upper river where it occurred. And there were dead fish eight or ten feet up in the mangroves," DeGaeta said. Outside their house, where the river was a third of a mile wide, the bodies extended from bank to bank.

After the spill, it took two years for fish to return. The seagrasses DeGaeta remembered from his first day on the river never came back at all. Earl Frye, the director of the Florida Game and Fresh Water Fish Commission, called the river "a solid mass of junk," telling *The St. Petersburg Times,* "It was practically a total kill."

Then, four years later, it happened again—the white spots, the dead fish, the sludge. Another spill, again from a phosphate mine in Polk County. Two billion gallons of slime entered the river from a Cities Service mining operation, devastating the water and killing 90 percent of the fish. This spill was worse than the last. ("Pollution control officials reported the only thing they can do at this point is test the water and watch thousands of fish struggle to the surface gasping for oxygen," Fred McCormack wrote for the *Sarasota Herald-Tribune* at the time.) The river became unusable, property values dropped, and Paul DeGaeta's family moved away. DeGaeta became a captain, piloting ships through rivers and seas, and then, in 1985, he married, returned, and took up teaching at the local high school. He bought a house seven doors down from where he had grown up. DeGaeta and his wife had three sons. They baptized each in the river.

Addressing DeSoto's board of commissioners in 2018, DeGaeta said, "Thank you, Mr. Chairman and Commission, for having your neighbors from the south be here. I'm not an expert witness, but rather I'm an eyewitness with four phosphate spills on the Peace River, including the two worst, 1967 and 1971. I experienced the stench of catastrophic fish kill caused by those phosphate spills. I experienced the decline in the living conditions and property values wrought on my small riverside community. My testimony provides an historic fact-based viewpoint to address several of the planning commission's points. There's an old saying that those who do not remember the past are condemned to repeat it." He began reciting newspaper headlines from phosphate spills and quoting the lackluster industry responses to them. In 1951, following a spill at a Virginia-Carolina mine in Florida, the manager said, "It was just one of those things you can't account for." Following a break a year later at a Swift & Company operation, the superintendent of mining operations "said he had no idea what caused the dam to break." There had been thirty major spills in the Peace River since 1935. DeGaeta presented a list of all of them to the board. "He was out there talking in his quarterdeck voice," Mele said of DeGaeta's presentation. Brooks Armstrong, who also spoke at the hearing, said, "He's got that booming voice. He's a schoolteacher, but he's also a captain. He's got a commanding presence when he's in the room. He's seen the history."

After the testimony finished, the board deliberated for thirty-six minutes, and Elton Langford, a board member, began to speak. Langford talked about the longevity of a phosphate mine, about the characteristics of Horse Creek, about the risks of mining and the need for phosphate around the world. "You could have heard a pin drop in the room," Mele told me of those moments. Langford motioned to deny Mosaic's rezoning request. Judy Schaefer seconded it. Terry Hill opposed. Jim Selph, the chairman, voted for it. Buddy Mansfield's name was called. Eight seconds passed in silence. Mansfield was a rodeo man—he had opened an arena that bore Mosaic's name. "Aye," he said. The board denied the rezoning, four to one. The room erupted. "Everybody was delirious," Paul DeGaeta told me. "It was a tremendous day." "To the best of my knowledge, it was the first hard stop that had ever been handed out to Mosaic," Mele said.

But Mosaic is a bigger organization than DeSoto County is, and the company threatened to sue the county after the vote. The two sides entered mediation, and the commissioners agreed to consider the request again in 2025. In 2020, Paul DeGaeta showed me the river where he grew up. By boat, the ride up the Peace from DeGaeta's house to Horse Creek is about ten miles. It starts out wide but narrows substantially. As we rode, DeGaeta talked about the movement of phosphate down the river in times gone by, and on the railroad that parallels it. The railroad used to take phosphate down along the Peace to Boca Grande, where a massive dock would load it onto ships bound for the world. "The gears, the creaking—it was like an old movie," he said. There were ruins of bridges and of towns, there were old wooden structures, but there was little current human habitation along the river. We passed a field of red mangroves and a row of blue crab traps, the top halves citrus orange, the bottoms clean white, lips around the middle. We reached the mouth of the Horse.

Horse Creek at Peace River is visually undisturbed by human hands. The creek evokes a sense of dark mystery and a feeling of wilderness. The trees are closing in on something deep. The animals inhabit, it appears, their own parallel world. "That's a primal stretch of the river," DeGaeta told me afterward. "To see the things we saw—where you can see alligators, where you can see turtles, where you can see a variety of birds. How many people get to see roseate spoonbills on a river? And it's just old, natural Florida, the way it's always been portrayed but seldom is. And it's peaceful. . . . Your soul feels better when you're on it."

The day we visited, no other people were around. The water was inky and quiet. The air was warm and still, and we rocked silently at the base of the creek. The future of phosphate mining in DeSoto remained uncertain.

From peak to valley, Barbara Fite has witnessed the history of an industry that, in Florida, is close to finished. She spent her early childhood in Brewster, a phosphate mining town. Her family had long been involved in the phosphate industry and had been present at various of its important moments. In 1930, Fite's grandfather, David M. Wright, the general superintendent of Swift & Company's Agricola mine in Polk County, was one of a trio of phosphate industry employees to file a patent on a process to separate fine grains of phosphate from quartz sand through the use of oil—the process that results in clay settling areas, which now are at the center of controversy in DeSoto. "When these men were researching this in the late 1920s, my mother still made mayonnaise at home," Wright's son, Hugh Wright, wrote in a 2001 edition of the *Polk County Historical Quarterly*. "Mayonnaise is an emulsion of vegetable oil, eggs, lemon juice and other ingredients, and I can remember these three men coming once to her kitchen to observe just how she blended the oil and other ingredients to make the emulsion. Vegetable oil was used in both processes."

Hugh Wright, David Wright's son, was Barbara Fite's father. Wright spent his whole career working as an engineer in the phosphate industry, and he and his wife, Freddie Wright, volunteered as local historians. She edited the *Polk County Historical Quarterly* for twenty years; he joined her as co-editor for the final eight. When Fite was forty years old, her parents interviewed Frank Garcia, a local legend in fossil hunting, and Fite joined them for the interview. (Garcia was, and is, renowned for uncovering more than two dozen new species in Florida's phosphate strata. Daryl Domning, the Smithsonian research associate who oversaw his collecting, told me that what Garcia found in those years "fills a half a long row of cabinets in the museum collection.") Soon afterward, a group of local fossil collectors, including Garcia, founded the Tampa Bay Fossil Club, and Fite became one of its first members.

More than sixty new species of prehistoric creatures have been discovered in Bone Valley over the past one hundred years, and most of those discoveries have originated in the phosphate mines. In 1990, Fite found a saber-toothed cat jawbone at an operation near Fort Meade, just south of Bartow. Fifteen years later, she found another one, this time at the Fort Green Mine, east of Bowling Green. She took the specimens to the Florida Museum of Natural History, where Richard Hulbert, a paleontologist and the vertebrate collections manager at the museum, identified it as a new species. Hulbert and Steven Wallace, a saber-toothed cat expert at East Tennessee State University, named the species *Rhizosmilodon fiteae* in her honor. Fite's contribution revealed a new species but also helped to clarify a longstanding question about the origin of saber-toothed cats—

the genus *Smilodon*—as a whole. "It's the oldest common ancestor known with the *Smilodon,* and that's one of the reasons that we gave it the scientific name that we did. We called it *Rhizosmilodon,* which means 'root of *Smilodon,*'" Hulbert told me.

Fite worked as a public-school guidance counselor, then retired and became a volunteer. The first time we met was at the Polk County History Center in Bartow, where she was dropping off an alligator skull for an exhibit. That was two days after Election Day, when she worked the polls, as she does every year. She manages dozens of vendors at the Tampa Bay Fossil Club's annual FossilFest event. ("She is the one that spends too many hours to count getting us a place for our show every year," someone posted in the online group Fossil Forum in 2011. "If you only knew how it is to deal with the Fairground officials you would hug her.") She drives a Prius that is cluttered with old papers and name tags and containers of various kinds and has a vanity plate that reads SBER CAT. "I couldn't believe it was available," she said as we drove along the oblong lakes and phosphate fields that make up Polk County.

Scattered across this landscape are square-shaped ponds, odd chemicals, and world-class golf. Something about the blank canvas of a disused strip mine has long appealed to those with an eye for golf course design. The Coronet Phosphate Company built a golf course on a Florida phosphate mine in 1929, and the International Agricultural Corporation built its own in 1937. With the land disrupted, the companies possessed the freedom to invent their own landscapes. They created hills and valleys wherever they pleased. In 2013, Mosaic unveiled Streamsong Resort, a luxury golf facility that covered sixteen thousand acres of mined land in Polk County, close to Brewster. The resort featured a sleek clubhouse overlooking a lake, with a spa on the lower levels and rooms upstairs whose price began at four hundred dollars a night. There were four restaurants constructed onsite. "I think what separates us is the all-in experience—the sense of the moment on the courses when you're playing, the unexpected topography matched with the unexpected architecture of our buildings," Richard Mack, the Mosaic executive who spearheaded the project, told the luxury lifestyle magazine *Robb Report* at the time.

Within an artificial landscape, Mosaic connected Streamsong to Florida's past. The company described building among dunes piled fifty years ago and made of sand that was ground up many millions of years before that. Rooms were furnished with *A Land Remembered,* a book that recounted the lives of early white settlers in the region. A rooftop bar was named after the poem "Fragmentary Blue," Robert Frost's celebration of the sky. ("Why make so much of fragmentary blue," Frost wrote, "When heaven presents in sheets the solid hue?")

The landscape of the golf course had been made by company towns, which, in the early days of mining, deconstructed wilderness. Pierce, Chicora, Agricola, Greenbay, and Pembroke all provided schooling, medical care, and food for families of miners, and they represented a different era of mining. "Times like that were so safe," said Patricia Prine, who lived in Brewster, Barbara Fite's hometown, beginning in 1938.

But the phosphate mines were all dug out, and the towns disappeared once they were no longer needed. In 1962, Brewster was dismantled, the land was excavated under it, and many residents moved to Bartow, the county seat. Prine was devastated. "I would sit there where the old depot was," she told me, "and I would cry, actually cry. I could not envision Brewster there. I said, 'Where did Brewster go?'"

Brewster is an oblong pond now, and, nearby, the Streamsong resort begins. "We just think it's hilarious that they made it into a golf course," Fite told me once when we visited the site together. The land stretched endlessly all around us, rolling hills pockmarked with small lakes, the residues of a landscape marred by mining. The sun lighted up the air far off, causing lakes to shimmer. The clouds were fluffy and floated in the sky. We could see into the distance without seeing anything. Far away, the land was fluid, but close up the ground was geometric, straight lines of sidewalk, grass, asphalt, and sand. Narrow tracks of gravel wound their way around grassy knolls. An osprey dipped down to catch a fish in a mining pond; failing, it returned to try again. Here, in the middle of a crowded state, atop a daunting hill, was a vastness so secure as to be nearly uninterrupted. It was a savannah; it looked wild; it was a manicured image of the wild. The horizon was flat, upended in small ways by distant trees. It was a different kind of dreamworld.

At Streamsong we found irony and history intertwined. The restaurant at the main lodge was called the P_2O_5, named after the formula for phosphorus pentoxide. At the end of one wing, the lodge displayed a history of phosphate mining. Within this setting, visitors were treated to images of the mining of prosperity over time. Photos showed the path that phosphate took, from the land of Polk County across to Tampa or down the river to Punta Gorda, as DeGaeta had described. One photo showed a rickety building on the river, with ladders leaning against it, a top that looked like a slatted wooden caboose, four stories of wood and pulleys, a dock, and two smokestacks. There were scraggly trees along the riverbank. Two barges together were almost as wide as the river itself. Below the photo was a plaque: "Founded in 1891 and one of the earliest mining companies established after the discovery of phosphate in central Florida, Bone Valley Phosphate is a direct corporate ancestor of The Mosaic Company, Streamsong's owner and developer."

Tied up in this extractive experiment are the lives that led to the formation of the rock. At Streamsong, those lives have been well venerated. Fossils are preserved in a large glass case—three rows of leg bones, jawbones, and teeth. Tapir, rhino, mastodon, dugong, horse. Sharks' teeth all in a row. Another plaque offered a history of the place, where animals roamed and golfers played. "Thick forests and grassy plains provided refuges for shovel-tusked mastodons, hornless rhinos, humpless camels and 30-foot crocodiles. Warm waters surrounding the area were filled with a rich variety of prehistoric marine life, including the notorious and now extinct giant shark, Megalodon." This land was traversed by indigenous peoples for twelve thousand years, then by Spanish explorers in 1539 and other European settlers after that. "Our aspiration—and inspiration—for this former mining site was to create a world-renowned luxury resort and golf destination, one that celebrates the heritage of the land," the plaque read, earnestly. "We hope your experience while visiting with us is as authentic and fulfilling as ours has been making Streamsong a reality for you." (Mosaic would sell Streamsong, for $160 million, in 2023.)

Fite was standing in front of the display cases. "Look at this," she said, motioning at a jawbone. "It's from a giant camel. They're hard to find. And I think that one—the teeth look really fake. The teeth are fake, I think. Actually, these might be replicas here. And that one—that one looks real. I think that might be a replica. These are real. Oh, this is a juvenile. The last tooth is still in the jaw." She was pointing at a succession of eight jaw pieces and a tusk, arrayed and labeled. The teeth of some were bigger than the jaws of others. A too-perfect jaw usually indicates a replica.

Behind us, through windows, the sun was nearing its horizon, sherbet orange reflected in the pond. The blue sky above was peppered with clouds. Fite stood by the fossils and spoke. "These are the little horses—tiny little horses. And this is the little tapir. And the bear—they call it bear dog because it was so big and could crunch up anything. And the giant camel, which is really big, too. And then the dugong and walrus, and the gomphothere and mastodon. They don't have a mammoth in here. I'm surprised they don't have some mammoth teeth. They have some lousy-looking sharks' teeth. You'd think they could do better than that."

III. REBIRTH

THE PACIFIC WORLD

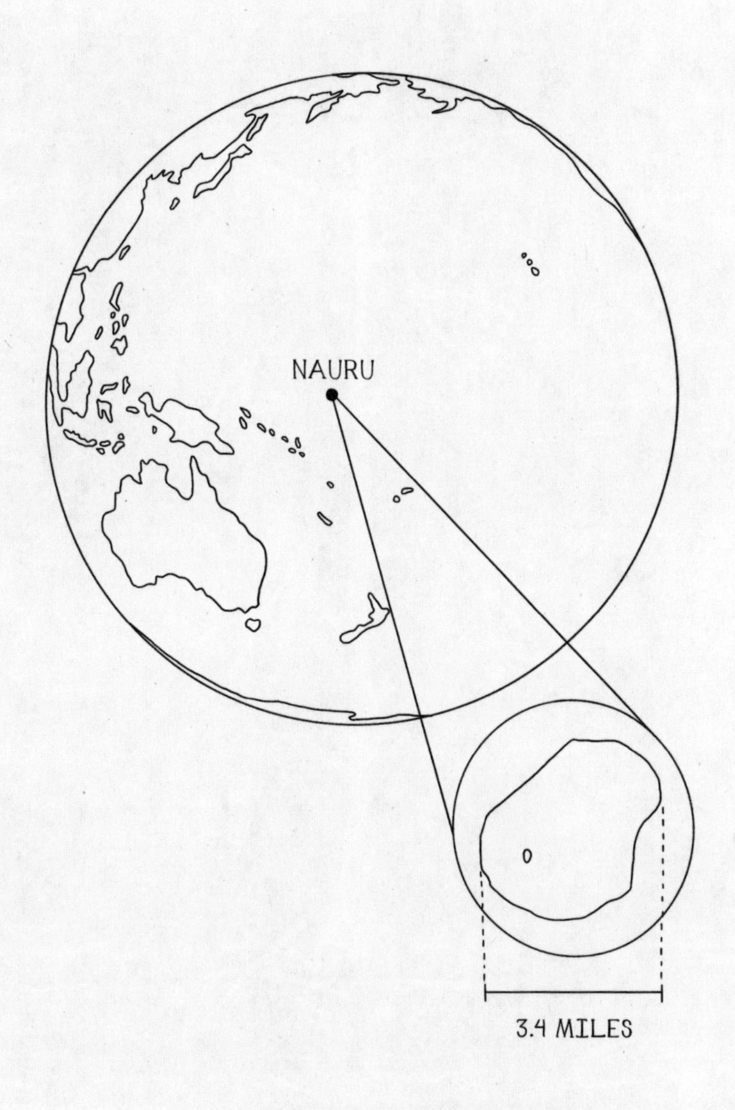

8

THE END OF EVERYTHING

An announcer was calling numbers in a national bingo game. She spoke into a microphone beneath a canopy, in the dark. "First number. One six, sixteen," she said. "Lucky seven. Double three, thirty-three. Seven zero, seventy." Most of the players sat inside parked cars, windows down, and the rest of us hovered in between. A player won; he honked his horn. A new round started. "Eyes down, good luck," the announcer said.

A world had fallen into ruin, and we were set within its midst. All around us, there were hints of past prosperity. Rusted machinery blocked our vision. Empty warehouses were covered in vines. There were, in the distance, crumbled concrete structures, abandoned buildings, iron trusses in the sea. The announcer's table was set up on a patch of ground that doubled as a playing field. Cars were parked in rows that filled the area, damp sand beneath their tires. I sat on a stack of metal beams; it was not uncomfortable. In a country of ten thousand people, hundreds were present at that night's game, gambling on prosperity. In this country—the Republic of Nauru—history is tragically intertwined with the lore of wealth; bingo is a remnant of that history. Nauru contains many such reminders. "It's a place where the world ended, and what exists afterward is this blind spot where things don't make sense," I had been told, before I arrived. "The world ended, and everybody stayed alive."

This is a place that once was known for fortune. A hundred and twenty years ago, a piece of rock was dug up here and transported to Sydney, where it came to prop a door. The businessman whose door it propped mined phosphate. He saw something familiar in his doorstop, and he sent to have it tested in a lab. The phosphate it contained made Florida's pebbles look like regular dirt. The twentieth century had just begun.

Nauru was inhabited. It was eight square miles, raised in the middle, covered in pandanus trees, and shaped by humans who had lived there for millennia. It was located midway in the Pacific between Hawai'i and Australia, a fleck of green on a sprawling sheet of blue. When miners came, the purest phosphate rock the world had seen was pulled out of the ground. An amount approaching 100 million tons of rock would be exported. In a century, the people of Nauru would become the richest in the world, and then close to the poorest. Four-fifths of the island's surface would be dismantled. Phosphate, arriving with promise, would leave behind a massive void. There would be no more land for farming, no more knowledge of how to live. There would be decayed machines, remaining bits of trains, pieces of draglines, trucks littering the island. By the end of the twentieth century, when all the mines had emptied, this place enriched by phosphate would collapse.

Here the island's story ends, and Nauru becomes a place outside of time. In the desperation of a destroyed land, a once-rich country replaced its mineral wealth with human lives. The mines became prisons. Asylum seekers from the world over, looking for a new life inside Australia, were shipped to Nauru and held against their will. Australia's government, enriched by Nauru's phosphate, subsidized their suffering. It sought to punish those who breached its borders. The money from the camps replaced the income from the mining, and the void that phosphate left had been replaced. Nauru was not as rich as it had been, but it survived. Of those who lived within the camps, the same could not be said.

To hide abuse within the camps, the government blocked all media from the country. It denied admittance to attorneys, advocates, journalists, and lawmakers. It deported its judiciary and arrested the opposition in Parliament. It banned Facebook. By 2018, Nauru had become a place turned upside down, a confusing mix of future and past. In the absence of everything—in the aftermath of phosphate—the island had become a kind of epilogue. The Republic of Nauru, a country beyond time, had disappeared. When these events were taking place, I arrived in Nauru and made my way to an old cinder-block hotel where I would stay for twenty-nine days. It was not a country I had ever expected to enter.

The promise of phosphate rock, like that of the philosopher's stone, was one of timelessness. The philosopher's stone was reputed to be a gift from God, and phosphate rock was often thought to be the same. ("He hid beneath the surface of the earth treasure houses of precious things which from time to time he brings forth for their welfare and provision," Leonard Jenyns had said in Bath.) But the philosopher's stone was never found, and phosphate rock turned out to leave a complex wake behind it. It had promised freedom from the cycles that govern life—cycles of the seasons, of animal lives, of the waste and replenishment that enrich the soil. In Nauru, this freedom meant that daily life lost its alignment with what surrounded it. Like the boats of ancient Edo that brought vegetables to the city and took the solid waste away, the ships that carried Nauru's rock could bring back food that had been grown with phosphate fertilizers. But unlike Edo's nutrient exchange, this new system did not renew itself, and it therefore had an end. In modern-day Japan, sewage recycling continues. Today, in Nauru, the rock is gone.

The phosphate company that mined Nauru imported huge steam shovels that towered above the trees, tearing apart the land. Phosphate was exported beginning in 1907. Hundreds of foreign workers, brought in from poorer countries and underpaid, provided most of the labor in the mines. A train carried phosphate from inland areas to the coast to be sorted, dried, and prepared for shipment. Albert Ellis, the Australian whose doorstop came from Nauru, went on to direct the first few decades of mining. He recounted the process in his 1935 book *Ocean Island and Nauru: Their Story*. When a ship arrived, Ellis wrote, a conveyor belt carried the phosphate from the coastal storage bins out over the coral reef by way of a giant metal swing-armed contraption designed to transfer the phosphate from the shore to the ship. Once out above the waves, the phosphate would be deposited in a 250-ton hopper suspended in the air. Then, "emerging from this hopper the stream of phosphate is divided, and two smaller conveyors take it up inclined galleries, where automatic weighing machines are installed, discharging at a point immediately over the centre of each swinging arm. There it drops on the conveyors running to the outer end of the arms as already described, and passing along the extension boom, it is discharged through a two-foot telescopic spout into the vessel's hold."

Huge amounts of phosphate left the island, and this period of time was characterized in Nauru by increasing extravagance and immense prosperity. At the height of mining, Nauru's government built a fifty-two-story high-rise in Melbourne and an eleven-hundred-home neighborhood in a suburb of Portland, Oregon. It founded an airline and bought a resort in Hawai'i and funded a London musical about Leonardo da Vinci. In the 1980s and 1990s, when its phosphate reserves began to diminish, the country's projects turned desperate: the government sued Australia for environmental destruction, sold passports and bank licenses, and became a global center for terrorist financing and money laundering. Because mining had destroyed the land, Nauru was unable to produce its own food, instead relying on imports from Australia. The country's institutions crumbled, fresh food ran out, and the national airline was reduced to one plane that sometimes failed to fly.

As Nauru collapsed, Australia grew. Ellis, identifying phosphate on Nauru in 1901, had written that the rock, mined and processed into fertilizer, would "make the desert blossom like the rose"—and in Australia, it did. Between 1913 and 1938, Australian consumption of superphosphate increased from 300,000 tons a year to well over 1 million. In a land whose nutrient-deficient ground had long stymied farming, the application of superphosphate to the soil resulted in multifold yield increases, and agricultural productivity became a source of economic strength and national pride. By 1979, Australia had, per capita, the fourth-highest rate of fertilizer consumption in the world, far outpacing the United States, the United Kingdom, and other countries that had pioneered the use of artificial fertilizers. That year, the quantity of fertilizers applied on farms was equal to 107 kilograms per person nationwide. Together with New Zealand, Australia also reached a milestone shared by no other region of the world: farmers' use of phosphate outpaced their use of any other fertilizer. In Asia, for example, more than 15 million tons of nitrogen fertilizers were applied annually, compared to a little over 5 million tons of phosphates. In Europe and North America, the ratio of nitrogen fertilizer to phosphate fertilizer applied was roughly two to one. In Africa and Latin America, the applications were equal. In Oceania, however, phosphate dwarfed nitrogen application by a factor of 4.5. And of the 44 million tons of rock phosphate brought into Australia between 1950 and 1980, half came from Nauru. By the end of the century, Australia's soils had been transformed.

Out of many, Nauru was only one. The islands were all scatterings and diffusions, pieces of rock that rose above the sea. Throughout the Pacific, phosphate was akin to political power. Up and down the west American coast, successions of empires collected guano, rich in phosphate, for millennia. It was a crucial material for the Aztec empire. The First People of Easter Island worshipped guano birds. Under the Inca, the status of the birds was carefully maintained. The system of guano extraction was efficiently organized. The mainland political system was replicated in miniature in the sea. The islands were shared between provinces and divided into village parcels in which families possessed shares. The global guano trade, which started in the nineteenth century, followed a path set by whalers. In 1828, an American seal hunter on the Namibian island of Ichaboe reported back on the presence of guano there, and fourteen years later, American whalers set out to harvest it. The Americans were joined by hunters from Europe, and half a million tons were taken from Ichaboe and its neighbors up and down the Namibian coast and sent to Britain alone, with other shipments going elsewhere. "British guano pirates headed next to Patagonia, then to the Khurīyā Murīyā Islands off the southeastern Arabian Peninsula," the historian Gregory Cushman wrote in 2013. The American Guano Company, formed in 1855, mined guano on a series of islands now part of Kiribati, near Nauru. The operation was supported by the whaling industry and staffed through the efforts of colonizers in Hawai'i. Governments around the world, from France to Mexico to Japan, collected overseas possessions on the basis of obtaining phosphate through guano. The Guano Islands Act, signed into law by President Franklin Pierce in 1856, enabled the U.S. military to seize and protect guano islands around the world, resulting in the collection of sixty-six islands by the U.S. government. Some islands controlled by the United States today, including Baker and Howland in the Pacific, were claimed under this law.

Phosphate money changed the course of history for Peru, where export of the substance had begun in 1840. Over the course of forty years, Peru exported 14 million tons of guano. It was shipped to Britain and to northern Europe, to Germany, the United States, Cuba, Puerto Rico, Mauritius, China, Brazil, and Australia. A quarter of all English farmers used guano during this time. The money made through guano sales—well in excess of £100 million—paid for the abolition of slavery in Peru and the end of an abusive tax levied on indigenous people. It funded the reconstruction of Lima and the new construction, in Peru's Andes, of what were then the two highest railways in the world. There were new urban sewers and new rural dams, a country reinvented via master plan. "In July 1872, vast crowds attended the opening of the centerpiece of this project, a Great Exposition modeled after European world's fairs. It left behind a vast, leafy park containing a zoo, an artificial lake, a formal French garden, and a permanent museum to art, science, and industry. A steam railway headed south from the park to the burgeoning beach resorts and cleansing water of Miraflores, Barranco, and Chorrillos," Cushman wrote. From 1879 until 1883, Chile fought a war against

Bolivia and Peru over control of the coastal guano reserves. The War of the Pacific stretched from the mountains to the sea and ended with Chilean territorial gain, although Peru maintained control of its main guano islands, the Chinchas, which continued to produce.

During the 1870s, the British businessman J. T. Arundel set about locating new islands while scraping clean the ones that had already been found. Phosphate samples were collected from Baker, Bat, Bunker, Caroline, Christmas, Clipperton, Ebon, Elizabeth, Enderbury, Fanning, Mole, Mouse, North Avon, Nukufetau, Passage, Solomon, Starbuck, Surprise, Walpole, and Washington Islands. High-grade phosphate was identified on many of them. On Fanning, phosphate readings were as high as 81 percent. On Washington, they were a point higher. In 1911, Arundel's Pacific Phosphate Company sent a representative to a collection of islands far southeast of Nauru—Nor, Aliar, Dito, and St. Augustine—but turned up only minimal phosphate. Walpole Island was mined by the Walpole Island Phosphates Company. Fanning and Washington were put up for auction in 1909. An advertisement for the sale of Washington Island read, in part, "Would you like to be a king, to buy an island, inhabitants and all, for your very own and to live free, truly monarch of all you survey in a balmy South Sea land with nothing to do but watch the cocoanuts grow?" The Pacific Phosphate Company acquired both islands three years later, estimating that 2.4 million tons of guano could be raised.

The Pelsaert Group was part of the Abrolhos Islands, a chain off the coast of Western Australia that contained a dozen atolls mined for phosphate. West Wallabi Island, Pelican Island, Pigeon Island, Rat Island, Beacon Island, and Gun Island were all among the earliest to be mined. Holbourne Island, in Queensland, also produced a significant amount of material. The mining spread across the Pacific. Ashmore Island was mined, Angaur in the Carolines, Christmas Island southwest of Java, Japanese Daitō, Fais Island in the North Pacific, Makatea—now part of French Polynesia—and, in the language of the day, Pelew, Rasa, Rota, Saipan, and Tocobe. In the years following World War II, phosphate on these islands was developed by British, French, and Japanese interests, which shipped the phosphate mainly to Australia, New Zealand, and Japan. During World War II, when

Axis forces captured Nauru, all the other islands filled a need. "With Nauru shot to pieces, the little island of Makatea provided nearly a million tons of phosphate between Pearl Harbour and V-J Day, the greater part of which came to New Zealand," *The Mirror* reported in 1946. "The armies which 'marched on their stomachs' to victory in all theatres of war, did so in good measure by the fruits of the soil of New Zealand and Australia—and the production of those fruits depended enormously upon phosphate supplies."

In Nauru, mining commenced in 1907. On Banaba, Nauru's closest neighbor, it had started seven years earlier. In history, quality, and scale, those two islands were distinct. Usually, guano differs from other phosphate deposits because it is neither rocky nor old. It may be years or decades old, or maybe older—but the guano harvested on islands does not exist on the temporal scale of rock deposits, whose ages rest in the range of millions, sometimes hundreds of millions, of years. The material found on Nauru and Banaba was an anomaly; it existed in a geological space somewhere between guano and phosphate rock. On those two islands, guano had been fossilized and melded with the remains of coral reefs. It was rock-like. Throughout the twentieth century, Nauru and Banaba would become the most productive guano islands of all. The patterns of the ocean gifted them the purest known phosphate in the world, and they suffered for it.

During the twentieth century, as phosphate disappeared from Nauru, the Oceanian writer Epeli Hau'ofa carefully chronicled the changing realities of the Pacific world. Hau'ofa was an international scholar, a Tongan born in Papua New Guinea who lived in Fiji until his death in 2009. "The latter part of the twentieth century has made it clear that ours is the only region in the world where certain kinds of experimentation and exploitation can be undertaken by powerful nations with minimum political repercussions to themselves," Hau'ofa wrote. Nauru—an island at the confluence of multiple kinds of experimentation and exploitation—was just one example of an experiment gone awry.

When cycles are supplanted by linear progressions, societies change. Hau'ofa wrote that Western people see chronology as a focal point for arranging facts, but Pacific cultures view time as a circle rather than a straight line. Hau'ofa described history as a spiral: a layering of environmental cycles that pull time from one year to the next by way of weather patterns, rainy seasons, and changes in the coral reef. The community's perception of its place in time is always dependent on the status of the land around it. As outside interests degraded islands and sea, the Oceanian perception of time that Hau'ofa describes was largely lost. Time, land, and memory used to be interchangeable elements that were connected by the actions of people and stories passed down over time. "Sea routes were mapped on chants," Hau'ofa wrote. "Distances were measured in how long it generally took to traverse them. . . . We cannot read our histories without knowing how to read our landscapes (and seascapes)." But imperialism and colonialism degraded and divided the whole of Oceania, breaking it up into constituent parts. Hau'ofa characterized Oceania as a "sea of islands," a large, interconnected space belying the perceptions of Western countries that see Nauru and its neighbors as small and remote. By turning these islands into countries and territories, he wrote, imperialists "erected boundaries that led to the contraction of Oceania, transforming a once boundless world into the Pacific Island states and territories that we know today."

In Hau'ofa's worldview, the decimation of land could not be separated from the subsequent loss of culture, governance, and human dignity. "It is the destruction of age-old rhythms of cyclical dramas that lock together familiar time, motion, and space," Hau'ofa wrote. "To remove a people from their ancestral, natural surroundings or vice versa—or to destroy their lands with mining, deforestation, bombing, large-scale industrial and urban developments, and the like—is to sever them not only from their traditional sources of livelihood but also, and much more importantly, from their ancestry, their history, their identity, and their ultimate claim for the legitimacy of their existence."

In 1938, in an attempt to preserve his country's past, Nauru's head chief, Timothy Detudamo, published a set of transcribed Nauruan lectures. The stories he wrote down embodied the peacefulness and egalitarianism that had characterized Nauruan society for three thousand years. One of the lectures recounted the story of Eigigu, a woman who grew dissatisfied with life on earth and, according to lore, planted a tree that grew and grew until it reached beyond the clouds. Eigigu climbed the tree and entered a new world full of danger and mystery. She outwitted a witch and wound up marrying the moon because she liked its quiet and cool and calm.

Detudamo fought a losing battle. Albert Ellis, in his own literary foray three years earlier, had described his experiences on Nauru. Ellis wrote that "the dream has been realized at Nauru, and strangely enough, the measure of success thus far attained has much exceeded even the dream—a pleasing state of affairs that does not often happen." A review in London's *International Affairs* praised Ellis's work. "The book is worthy of its author, modestly and competently written," the reviewer wrote. "The story of Ocean Island and Nauru will not need to be written again."

The white settlement of Australia began as a prison colony, a 1787 shipment of eight hundred convicts sentenced to live out their years on an unknown landmass ten thousand miles from home; their role was to be punished and to serve as an example for criminals in England. Watkin Tench, a marine who chronicled the colony's first few years of existence, wrote that the convicts' "constant language was an apprehension of the impracticability of returning home, the dread of a sickly passage, and the fearful prospect of a distant and barbarous country." Upon arriving in Australia, Tench described the convicts' responses to the place, including one letter from a convict who wrote home that Australia was "nothing better than a Place of banishment for the Outcasts of Society." Tench himself noted, "If only a receptacle for convicts be intended, this place stands unequalled from the situation, extent, and nature of the country."

A man guilty of petty theft "was sent to a small barren island, and kept there on bread and water only, for a week," while two other thieving convicts "were ordered to be kept close prisoners, until an eligible place to banish them to could be fixed on." Over time, the colony's leadership established a collection of decrepit island prisons where convicts were held, tortured, and sometimes killed. Among the worst of these was Norfolk Island, a rocky, formerly uninhabited space about a thousand miles from the Australian coast. Convicts were held on Norfolk Island from 1788 until 1814 and again beginning in 1825, when the colony reopened with a new remit of cruelty. For thirty years, it operated under a succession of despotic leaders who ruled the lives of the thousand convicts held there, administering harsh punishments and enforcing a state of fear. Hopeless, the prisoners turned "suicide into an act of solidarity, not of solipsism and despair," Robert Hughes wrote in a 1986 history: groups of convicts would "choose two men by drawing straws: one to die, the other to kill him." The former escaped by dying, the latter with a criminal trial in Sydney and, in all likelihood, death himself. This kind of disturbing practice, Hughes wrote, was an intentional by-product of a system that aimed to terrorize its subjects. "Norfolk Island held a thousand convicts, but its real use was the intimidation of tens of thousands more. If it was not 'demonic,' it would have been as useless a deterrent as a gallows with no rope. Mercy on the mainland needed the background of terror elsewhere."

The system of convict banishment overseas, which the British public had eventually begun to associate with slavery, ended in 1868. In total, the government sent 164,000 convicts to Australia. Many of them were poor agricultural laborers, some of whom had participated in the protests of the 1840s, lighting fires in Suffolk and beyond. Three of them, accused of stealing sheep, had been sentenced by John Stevens Henslow himself, who served not just as rector but also as the magistrate of Hitcham. Their desperate crimes became the basis of the country.

During the twentieth century, Australia developed a unique identity as an island-country-continent located far from Europe and the Americas: a place of prosperity, unburdened by the social ills that bedeviled the rest of the world. In maintaining its sense of isolation, Australia's government worked to keep nonwhite people outside the country's borders. After gaining independence in 1901, one of the first acts of Australia's Parliament had been to pass the Immigration Restriction Act of 1901, a statute that denied certain racial groups access to the country. That began an era of policy known as White Australia, defined by a collection of laws and regulations issued by the government that prohibited, for more than seventy years, the immigration of nonwhite people to Australia. The White Australia policy did not end until 1973.

After White Australia, the country's immigration restrictions continued in altered form. Parliament had legalized the discretionary detention of asylum seekers in 1958, while instituting a requirement that people be held for no longer than 273 days. In 1992, that time limit was lifted, and the detention of asylum seekers was made mandatory. "If an officer knows or reasonably suspects that a person in the migration zone is an unlawful non-citizen, the officer must detain the person," the new Migration Reform Act stated. In the years that followed, the government set up detention centers in Melbourne, Sydney, and Perth, a system that soon expanded seaward.

As Nauru's phosphate reserves ran out by the 1990s, the island faded from Australian consciousness. In 2001, however, when various groups of asylum seekers attempted to reach Australia by boat during an election year, Australia's then prime minister, John Howard, reached back into history and plucked two policies—deterrence and island imprisonment—to demonstrate to the public that he was trying to stem the flow of refugees toward the country. When in need, Australia remembered Nauru. Howard's policy, known as the Pacific Solution, consigned the boat arrivals to Nauru and Manus, two far-flung islands where the migrants would be out of sight and out of mind—but where news of their travails would reach places like Iraq, Afghanistan, and Pakistan, scaring other would-be asylum seekers from attempting the journey to Australia. The policy lasted until 2008, when the government closed the camps. Four years later, facing new election pressures, Australia's government came to Nauru with an offer of money and economic development, and the two countries agreed to reopen the camps.

One of the first people to arrive was a woman named Sadika. Sadika was from Kuwait, a country where she had never been safe. In 2009, her ex-husband had found her, tried to hit her with his car, and told her, "I can cut you into pieces and throw you in the desert and no one will miss you—because you have no one." Sadika had lived in Kuwait all her life but could not protect herself from her ex-husband's threats because, as a member of the Bedoon social class, she was denied citizenship by Kuwait's Arab government. Born in 1974, she married as a teenager and gave birth to two daughters before her eighteenth birthday. She later divorced but lacked legal standing to seek custody of her daughters. Sadika remarried after the divorce, but her second husband beat and raped her repeatedly over the course of their five-year marriage. "I have cigarettes he put on my hands, just to say, 'I love you,'" Sadika told me. As a stateless noncitizen, she lacked the right to protect herself.

Four years after her ex-husband tried to kill her with his car, Sadika left Kuwait. (She asked to be identified here only by her first name.) With the help of smugglers, she traveled first to Qatar, then to Malaysia, and finally to Indonesia, where she waited for more than three months before the smugglers placed her on a boat to Christmas Island, an Australian territory. She remembers feeling dizzy during that three-day journey across the open ocean, crammed with other people into a tiny boat not equipped for the sea. During the trip, she would wake up intermittently to find fish on her face; at some point, the boat began to sink. "The water, slowly, slowly, was coming up from where we sat," she said. People around her bailed out the water, and eventually the boat reached Christmas Island. Sadika was picked up on the beach and held in Australia's Christmas Island detention center for a short period of time before the government transferred her to Nauru, more than four thousand miles away.

Nauru presented a promise to Sadika, but that promise was not fulfilled. One month after arriving, Sadika was assaulted by a male nurse who worked for International Health and Medical Services, a company contracted by the Australian government to provide health care in Nauru. Sadika said the IHMS nurse assaulted her multiple times between November 2013 and September 2014. Traumatized by a lifetime of abuse, Sadika was already afraid of men, and she froze whenever the nurse touched her or forced her to perform sexual acts. "I didn't have any reaction," she told me. "I was scared to do anything." Kathy Barazandeh, an Iranian woman who lived in the camp at the time, shared a tent with Sadika and six other women inside Regional Processing Centre 3—RPC3, the family camp—and said that Sadika's mental health deteriorated after she was assaulted. "She got worse every day on the island," Barazandeh said. Nigel James, who worked for Save the Children Australia in 2013 and 2014 and knew Sadika, told me that the environment in RPC3 was uniquely hostile to women—a charge that has been documented by numerous government investigations and whistleblower accounts. "She did very well to manage. She got on very well with pretty much everybody. She was everyone's friend," James said. "But she was a single woman and she felt vulnerable. There were a lot of inappropriate references and sexist comments" made by guards at the camp.

On October 30, 2014, Sadika reported the nurse's assaults to the Moss Review, an Australian government investigation into allegations of improper conduct inside the Nauru detention centers. One month later, she reported the abuse to IHMS, submitting an official report. "I told them all the story," she told me, "and after a couple of months they came and said, 'We believe you, but we can't do anything. Because no one saw—no cameras, nothing. We can't.'" She said the people she spoke with at IHMS promised to supervise the nurse whenever he treated women in the camp. (IHMS, a subsidiary of the global travel consultancy International SOS, did not respond to several requests for comment about this story.)

In 2015, Sadika gained refugee status and moved into a community settlement. She painted the ceiling of her new room to look like a rainbow, but, haunted by her experiences, she still struggled to sleep at night. Her life became a blur of fear, self-harm, and thoughts of suicide. "Somebody said, 'Sadika had a bad experience with sexual harassment when she was in the closed camp,'" Barazandeh told me. "After that she got crazy and her mentality got worse every day." During this time, Sadika shared her experiences with mental health counselors from Overseas Services to Survivors of Torture and Trauma, an Australian nongovernmental organization contracted by the Australian government to provide mental health care in Nauru. Sadika met with counselors thirteen times between July and December 2015. Based on these meetings, according to a psychosocial report dated December 12, 2015, OSSTT determined that Sadika's account of the abuse was "highly credible." The OSSTT report described her as feeling vulnerable and depressed. "The incidents of indecent assault and harassment," it stated, "are likely to have exacerbated post-trauma symptoms and have resulted in her feeling trapped in a state of anticipatory fear."

Sadika had earlier reported her abuse to the Australian Border Force and been instructed to talk to the Nauruan police instead. On March 9, 2016, she submitted a formal police statement recounting six incidents of abuse by the IHMS nurse. She wrote that the abuse included "kissing me, hugging me, touching my breast, making me touch his penis and touching my bottom . . . and this nurse is still working for IHMS at the moment." Two months later, Sadika reported her allegations to the Commonwealth Ombudsman of Australia and provided copies of her prior complaints: a letter confirming her participation in the Moss Review, her complaint to the ABF, the OSSTT psychosocial assessment, and her statement to the Nauruan police. "To my knowledge," she wrote in the complaint, her abuser "is still working at IHMS Clinic, Regional Processing Centre one." It had been over eighteen months since she first reported her abuse, and thirty months since the abuse had begun.

At the end of 2016, when the United States signed an agreement with Australia to resettle 1,250 refugees from Nauru and Manus Island, Sadika, along with many others in Nauru, applied. In late 2017, Sadika's application was accepted. During her last days in Nauru, Sadika said goodbye to the people she had lived among and struggled with since arriving. Many people had gone to her with their problems, and she had listened. On the advice of a friend, she gathered documentation of her assaults and subsequent complaints, as well as other evidence of her time in Nauru. "Write whatever you have and have some pictures," the friend had told her. "One day you'll move, you'll go anywhere, and you can write a book—your story, your life."

In the United States, Sadika's psychological problems have persisted. She takes antipsychotic medications that make her feel tired most of the time. She is overwhelmed by crowded places like shopping malls. Wherever she goes, the sight of unknown men scares her. Sometimes she hears voices in her head. "Kill yourself," they tell her. In April 2018, she was admitted to the hospital as an emergency-status patient; according to a document prepared by the New York State Mental Hygiene Legal Service, doctors worried that she might hurt herself if they left her alone. In July 2018, she underwent a mental evaluation with county social services to determine her ability to work. The social worker who completed the evaluation noted that Sadika had "severe depression and psychosis" and had been prescribed daily dosages of fluoxetine and quetiapine, psychiatric medications used to treat depression. In a section marked "Employability," a social worker checked the box next to a statement that read, "Individual is unable to participate in any activities except treatment or rehabilitation." Sadika was prohibited from seeking employment for at least four months.

Sadika told me that in Kuwait, she felt like a normal person with a bad life. Her four years in Nauru took that sense of being normal away from her. When I visited Sadika at her home in New York, she had lived in the United States for one year and was trying to feel better again. She described weekly visits to the hospital and monthly trips, with a friend, into the community near where she lives. On a coffee table in her living room, she spread out the papers that catalog her years in Nauru—her painstaking documentation of assaults, cover-ups, and institutional apathy. A piece of Nauru lives inside her. "I'm crying and at the same time laughing and at the same time thinking a lot of mixed things. I talk with you, but my mind is thinking I'm there now, in Nauru. When we were there, we tried to show that we were strong. 'Oh, it's okay. It's better than nothing. We did not die in the boat. We are alive.' We lied to ourselves to continue life. But we were dying there. We went crazy there. We were not normal."

Sadika's experiences in Nauru were not unique. Since 2012, powerful pieces of journalism have alleged assaults and other kinds of abuse inside the camps. The Moss Review, the investigation to which Sadika first reported her abuse, supported these accounts. When the Moss Review was made public in March 2015, it documented scores of disturbing abuse allegations and revealed that two IHMS employees had been accused of assault. When an interviewer asked Australia's prime minister, Tony Abbott, about the findings, Abbott said, "Occasionally, I dare say, things happen, because in any institution you get things that occasionally aren't perfect."

Australia achieved independence 113 years after convicts first made the journey to its shores, and mining began on Nauru six years after that. For the majority of Australia's history as an independent nation, its soils have been fed by Nauruan rock. For the majority of its colonial history prior to independence, Australia was primarily a penal settlement that operated prison islands offshore. In operating refugee camps in Nauru, the country has returned to its roots.

The conditions that led Australia to establish camps in Nauru were similar to forces that affect the world today. The fear of outsiders, the corruption of leaders, and the distraction of a populace harried by the pressures of a late-capitalist economy all contributed to the decision to banish a collection of hopeful people to a depressing and wildly expensive fate. The flow of migrants to Australia in 2001 was a political crisis that demanded a political response—not an actual crisis that required a solution. The number of people coming to the country by boat that year was small compared to the number of migrants arriving, for example, by plane, but the drama of a boat arrival necessitated a response of comparable narrative weight.

From the beginning, the camps have been opposed by the portion of the populace that does not believe in the imprisonment of refugees in Nauru, whether because of the abuse that is perpetrated there or the costs the camps have incurred in other ways. Coalitions of activists publicized the plight of children in particular, leading to the evacuation of many young people in 2018. But the detention system in Nauru was supported by people who feared outsiders, by politicians who knew how to exploit that fear, and by those who realized they could profit from this infrastructure of abuse.

The governments of Nauru and Australia, and the myriad contractors they employ, typically direct attention at each other, rather than responding directly, when problems arise relating to the camps. Those who have made money on the island include IHMS and Canstruct, an Australian civil engineering firm that was awarded contracts totaling billions of dollars to manage the facilities. Reached by email in advance of this book's publication, a representative for Canstruct asked that the company be referred to as "an Australian Multi-Disciplinary Project Delivery Company" but declined to comment on abuses carried out under its management. Broadspectrum, which managed the camps before Canstruct did, has since been acquired by the Spanish multinational Ferrovial. In 2017, following an Amnesty International report titled "Ferrovial Continues to Build a Fortune on Refugees' Despair," Ferrovial argued that Amnesty's work was "gratuitous" because the Nauru camps were "not part of" Ferrovial's "strategic business portfolio." "The company considers that Broadspectrum is being held responsible for a range of matters well outside its scope," Ferrovial said. In 2022, Management & Training Corporation, a U.S.-based private prison operator, took over management of the Nauru facility. MTC has long attracted controversy. Among many allegations from many people about many MTC facilities worldwide, the American Civil Liberties Union wrote in 2018 that at the MTC-operated East Mississippi Correctional Facility, where mentally ill patients had died by suicide, "even the most troubled patients are routinely ignored and the worst cases of self-harm are treated with certain neglect." In 2022, when the company's Australian arm, MTC Australia, accepted the contract to manage Nauru's facilities, *The Guardian* noted that the cost of the program was 4.3 million Australian dollars per person, per year. MTC did not respond to a request for comment for this book.

Provided with details from this chapter, including Sadika's allegations, a representative of Australia's Department of Home Affairs wrote, "The Government of Nauru is responsible for the implementation of regional processing arrangements in Nauru, including the management of individuals under those arrangements." The government of Nauru did not respond.

Of all of the outside organizations that have been involved with refugee processing in Nauru, only Save the Children Australia provided a detailed response for this book. "Our dedicated staff, social workers and teachers worked tirelessly to protect the children transferred to Nauru, using all channels available to them to ensure children's safety and wellbeing. Their expertise and experience informed multiple Save the Children Australia submissions to official inquiries," a statement from the organization read, in part. "We consistently and vigorously raised issues directly with responsible Ministers in the Australian Government, the bureaucracy and at an operational level with service providers in Nauru. We believe this advocacy was a key reason why our contract for services was not renewed by the Australian Government."

Nigel James, the Save the Children Australia employee who knew Sadika, told me that he and other workers in Nauru were incentivized to bend rules in order to curry favor with the Australian government. James, a personal trainer, operated a gym based in RPC1, which held administrative offices and group facilities. At the time, Save the Children Australia held the contract to manage RPC3, the family camp, and Transfield managed the single men's camp, RPC2. "There was a lot of money involved, so stakeholders would tick all the boxes and follow suit with whatever the government wants," James said. "Otherwise, they may lose their contract." James and a half dozen other SCA employees oversaw activities for parents, married couples, and single women asylum seekers who arrived by bus each day from the family camp. The SCA workers' job, James said, was "to keep morale up" by holding movie nights and group cooking sessions, planning outings and organizing group sports such as soccer, football, and volleyball. When disagreements arose over what music to play in the gym, James would give everybody a shot at picking tunes. "People played their own music," he said. "We played a bit of everything. I liked to play a bit of Johnny Cash, myself. I'd say, 'This is Johnny Cash, good man!'"

Early on, though, some things felt off about the Nauru camps. James was responsible for chaperoning "excursions," guarded trips that allowed asylum seekers to see the beach in one part of Nauru. Guards and SCA personnel would ride the bus with a small number of people to a secluded area of oceanfront. Asylum seekers would be allowed to play in the sand and wade in the water, but not to swim or take any photos. Before getting back on the bus, each person would be searched, James said, "to make sure they weren't taking seashells back, in case they would've cut their wrists." The camp, James said, felt like a prison. The Nauru RPCs operated under a system of points similar to policies in Australia's onshore prison system. By attending an event—an English class, a volleyball game, a movie—people could earn points redeemable at a small camp store that sold snack foods, tobacco, and other products. Before entering recreational spaces, including the computer room and the gym, people were supposed to line up and be counted. They were forbidden from taking pieces of fruit from the cafeteria to eat later.

In response to these restrictions, people rebelled in small but meaningful ways by sneaking food out of the cafeteria and asking SCA teachers to smuggle in books for their children. "It was a bit like *Hogan's Heroes,*" James told me. "This is all the confusing, bizarre environment that Australian workers found themselves in." When issues came up around the camp—physical abuse, improper standards of care, leering statements by guards or other employees directed at women—James felt powerless to do anything. "You could write an incident report," he said. "You could pass that on to your superior, your coordinator. And they would probably pass that on to their supervisor, and then it might go on to another government office or department that might deal with it. But it might come back that you're complaining, or it could come back that you're not suited for the job, because you're too structurally obsessed with the rules. In this job, you can't go by the rules."

In 2015 *The Guardian* published over two thousand of those incident reports, which had been leaked from inside the system. The documents showed definitively the scale of abuse alleged to have taken place inside the Nauru RPCs: assaults, beatings, suicide attempts, and acts of self-harm, many of them involving children. The documents were all the more striking because Australia's government had previously threatened whistleblowers like James with two years' imprisonment, restricting the flow of information from the island.

The Nauruan minister for justice who approved the camps on the island was a man named Roland Kun. Slight and soft-spoken, Kun had risen to power in the void that followed phosphate, and he had tried to make the island whole again. He saw the camps as an opportunity to diversify Nauru's economy and help fix the social issues that plagued the island. His colleagues agreed. "Australia came with this opportunity, and there was no resistance," Kun told me. "We went straight into discussing the workings." Beginning in September 2012, the sudden influx of spending transformed Nauru, evoking memories of how mining had changed the island a hundred years before. Unemployment plummeted as the camps hired hundreds of Nauruans to work in management positions and as cooks, janitors, and security guards. Revenues increased: the Nauru government's budget grew from 27 million Australian dollars in the 2010–2011 fiscal year to 60 million dollars in 2011–2012, then to 99 million dollars in 2013–2014, and it kept growing after that. The physical landscape changed, too, as Australia pledged millions of additional dollars in infrastructure projects to build more housing in Nauru and rebuild public facilities, including the country's hospital. Soon cars filled the roads as years of hardship faded away.

When Nauru held parliamentary elections in June 2013, Kun won reelection to his seat. In the days that followed, he stood for president—a position selected by a vote of Parliament—but lost, thirteen to five, to an MP named Baron Waqa. Early on in Waqa's presidency, as the government deepened its reliance on Australian money, problems began to surface. A riot and other events revealed a chaotic atmosphere inside the asylum-seeker camps, which were located in the old phosphate mines at the center of the island. The U.N. Refugee Agency had already worried that the uncertainty of the process was "likely to have a significant and detrimental impact on the mental and physical health of asylum-seekers." Alarmed by these reports, Geoffrey Eames, the chief justice of Nauru's Supreme Court, and Peter Law, the resident magistrate, toured the facilities in November 2013. "There were kids sitting alongside the tents, because it was all just stone on the ground, and it was all just radiating heat," Eames told me later. "As the sun moved around, there'd be a slant of shade coming down the side of the tents."

The two judges were appalled by the conditions that they witnessed, and Eames spoke with Nauru's new president about their concerns. At that point, Eames and Law had been in Nauru for nearly three years, during which time they had worked to reform the country's judiciary, setting up a probation program, sending employees overseas for training, and creating a mediation option for civil disputes. Several weeks after Eames spoke with Waqa about the camp visit, Parliament passed legislation making it easier for Waqa's new justice minister to deport noncitizens. Soon after, the minister attempted to deport several local businessmen. When Law issued injunctions halting two of the deportations, Waqa fired Law. The next day, the new justice minister ordered his immediate deportation from the island. "I got a phone call from Peter one Sunday saying, 'I'm in a police car. I've just been arrested, and I'm being taken to the airport,'" Eames told me. "I was pretty astonished. It's not every Sunday that you remove a resident magistrate." Eames issued injunctions preventing Law's removal from the island and bought a ticket to fly to Nauru that night. At the Nauru airport, police were presented with Eames's injunction preventing Law's deportation. An officer told Law, "I don't take orders from the chief justice," and they placed Law on a plane to Australia.

Eames never made it to Nauru. Before leaving home, he was informed that his visa had been revoked; for two months afterward, officials from Australia and New Zealand negotiated with the government to let Eames return to Nauru. When those negotiations failed, Eames resigned as chief justice, issuing a statement that called the events leading up to his departure "an abuse of the rule of law and a denial of the independence of the judiciary." In his statement, Eames lamented the loss of the reforms that he and Law had championed. He expressed gratitude to the legal practitioners of Nauru, "whose courage and dedication will provide the best safeguard of the rule of law in future."

After Eames resigned, Roland Kun and two other MPs spoke to international media and condemned the ouster of the two judges. Kun called the government "either completely ignorant of the importance of the separation of powers and judicial independence—and therefore unfit to govern—or so contemptuous of the rule of law that they think they are entitled to effectively dictate to the judiciary that judicial decisions must favour government." In response, the government suspended Kun and his two colleagues from Parliament. Two more MPs were later suspended for defending the initial three.

Following his suspension, Kun and his wife, an Australian lawyer, moved to Melbourne with their three children. For six months, they lived in friends' homes; Kun's wife worked while he took care of their children. In mid-2015, Kun traveled back to Nauru for four days to meet with officials and discuss his suspension. Afterward, when he boarded a plane to return to Australia, police arrived and removed him, confiscating his passport so that he would be unable to leave the country again. For more than a year, the government detained Kun in Nauru; he was unable to leave or see his children. In response, the government of New Zealand, which funded Nauru's judiciary, withdrew its financial support. Eames spoke publicly about the Nauru government's "outrageous conduct" in its treatment of Kun. Lawyers' groups and other organizations condemned Kun's detention.

Kun's family moved to New Zealand, where his wife secretly began efforts to obtain citizenship for Kun on humanitarian grounds; with a passport, he would be able to leave Nauru. Thirteen months after Kun was first detained in Nauru, the citizenship process succeeded, and in July 2016 a friend delivered Kun's emergency New Zealand passport to his residence in Nauru. Aware that he might be under surveillance, Kun sent a coded message to his wife. "I'm going to have lunch with a friend on Sunday," he wrote.

That Sunday—the morning after the 2016 parliamentary elections, when the winners would be celebrating and unaware—Kun rode to the airport, checked in at the counter, passed through security, boarded the plane, and flew to Australia.

"MRC BINGO TONIGHT 8:30PM TIGERS OVAL—7 x $200, 1 x $1000, 2 x BRAND NEW YAMAHA CRYPTON MOTOR BIKE. DON'T MISS YOUR CHANCE ON TWERKY TUESDAY ONLY AT $20 BKLT" was the text message that I followed, one particular July evening, to Tiger's Oval, the field where the announcer called numbers at the heart of Nauru. The message was one of many, sent daily across the country, announcing the prizes for upcoming bingo games. Bingo occupies significant space in the modern culture of Nauru. Out of a population of ten thousand, hundreds could be found on any night, blotting on booklets, hoping to strike it rich again.

The equipment around the playing field was not entirely out of use. RONPhos—the Republic of Nauru Phosphate Company—was scraping the remaining bits of phosphate from the landscape. A train used to bring phosphate rock from the center of the island to the space near where we sat; today, trucks do the moving. When a ship arrives, wormlike conveyor belts move the processed phosphate along the side of Tiger's Oval, tunneling under the island ring road and emerging at the ocean's edge. There, huge swinging booms cantilever out over the coral reef and deposit the phosphate into the hold of a waiting ship. Many years ago, when ships arrived more frequently, four booms were used, two sets of two. The original pair, the booms closest to Tiger's Oval, have long since been abandoned to the surf, where they remain broken and visible.

A straight line runs through the history of Nauru, from the rise of phosphate to the loss of knowledge, from the loss of knowledge to the end of hope. In Australia, the cycles disappeared, too, and the culture suffered heavily for it. It became the home of White Australia, of the Pacific Solution, and of Operation Sovereign Borders, which oversees the Nauru camps today. Phosphate is the mineral that pulls humanity earthward, but phosphate mining has severed that connection. The easy promises of phosphate, like the early promise of the philosopher's stone, could not be kept. Instead, there are continual reverberations of the booms and busts that phosphate rock creates. When a mine is emptied, a world is gone; another is created somewhere else. The timelessness of phosphate came from bodies that lived long ago in a process that once governed the world. In the aftermath, there was a void.

Today, the heart of Nauru is an emptiness where phosphate used to be. A main ring road encircles the country; gravel roads branch off at several points and take motorists to the interior of the island, called Topside, which sometimes begins just a few hundred feet from the coast. The transition is sudden, from trees and houses to a hot, dry, and dusty landscape. Only a handful of roads traverse this area. All around, spreading out into the distance, are endless fields of pinnacles ranging in size from ten to eighty feet tall. They are the remnants of a coral reef that developed millions of years ago atop an undersea volcano, which, millions of years before that, had begun to thrust upward from the ocean floor, pushing the island out above the waves. Marine organisms decomposed around the bones of the uplifted reef, beginning the long process of phosphate formation. Today the exposed pinnacles populate the landscape like giant, faceless replicas of Emperor Qin's terra-cotta army. They are interrupted by several landmarks: a solar power field, a trash dump, two old Japanese guns, a Japanese prison, a Nauruan prison—funded by Australia but still, when I visited, unused—as well as phosphate mining equipment, new and old, and RPCs 1, 2, and 3. Over time, as people gained refugee status and moved into the coastal settlements, the RPCs have slowly emptied out, and RPC3 closed shortly after my visit. Most refugees then lived in community settlements that sat along Nauru's main road and had names like Anibare Lodge, Anabar Pond, and Budapest Hotel. When I visited, they already looked old.

Grand buildings rise up in some corners of the island—concrete hulks, massive structures built with mining largesse that give the island a distinct twentieth-century feel, like a decayed shopping mall.

When the mining ended, many buildings—including the two hotels and several mansions—fell into disrepair. Since 2012, activity has picked up again, and a new set of unsustainable structures has been built: cheap metal and plastic prefabrications that constitute much of Australia's construction on the island. According to data from the U.N.'s Comtrade database, prefabricated buildings were Nauru's number one import for the five years ending in 2016.

I met with Eames and Law, the two judges, in Melbourne on January 18, 2019, one day short of the five-year anniversary of the day Nauru's government arrested Law and revoked Eames's visa. "The fact that it's been five years is a sad commentary, because it means that this government has effectively gotten away with it," Eames told me. "It's been at great cost to a lot of people who've lost their jobs, and at great cost to the rule of law." Law said, "I think generally you could say that since what happened at the time of the termination and demise of our roles in Nauru, that the justice sector has been compromised. I think that's a general observation that's been confirmed by what's taken place since."

In Nauru, history and modernity intersect. "What happened to us in our heydays of phosphate mining and high economic activity due to phosphate mining, that's now replaying itself, but now it's with the backing of the offshore processing center," Roland Kun told me. He believes the only way to restore a functioning democracy in Nauru is for Australia to remove the refugee camps from the island, which would take away the Nauruan government's power to placate voters with money. "They know what wins votes, and that's what they're banking on—until the money dries up, unfortunately. Then the voters will start going, 'Well, why did you do that?'" Kun said. "After the offshore processing center exits, the money will dry up overnight." (In 2023, after ten years of imprisonment, the remaining refugees were removed from Nauru, although Australia's offshore detention policy persisted: the government continued to spend about a quarter of a billion dollars annually maintaining the empty detention facilities for future use. "Regional processing arrangements with Nauru are enduring and remain in place to receive new unauthorised maritime arrivals," the Home Affairs representative told me in 2024. Shortly afterward, people were sent to the island once again.)

Kun yearns for the day when a new wave of reformers rises up to fix the problems of Nauru, much as Nauru First, the party that he founded, tried to change the country's circumstances twenty years ago. The actions of the current government—their treatment of him, of the judges, of the refugees held on the island—have disappointed him. But he still feels proud to be a Nauruan. At night he tells his children stories from Nauru, including the story of Eigigu, transcribed by Detudamo all those years ago. Kun told me the story of Eigigu, too: how she felt dissatisfied on earth, how she planted the tree and escaped into the sky, how she took up residence in the moon. Kun tells his children that if they look up at the moon, they can see Eigigu as she watches down on the events of the earth she left behind. Kun told me all of this, and then he paused and said, "There are times when I'm telling the story and the details seem to blend too closely to 'Jack and the Beanstalk.' And the kids are going, 'You're just copying "Jack and the Beanstalk."' Sounds a little bit like I am, doesn't it? But I'm not, really. This is the story as it is."

9

OVERHAUL

On North Wyke estate, near the southwest tip of Britain, the cycles of the living and the dead are monitored on an hourly basis. The transfer of gases, the movement of water, and the conditions of the ground all undergo constant measurement, carried out by sixty researchers studying eight hundred acres of gently rolling hills. The topsoil of the hills is twelve inches thick and is girded by a bed of impermeable clay; water from the surface sinks into the soil, trickles down, and hits it. The water runs downhill along that clay surface, and it is captured, piped, measured, analyzed, and then let go. Grazing cows wear collars on their necks that measure how much methane they emit. Researchers are interested in the grass the animals eat, the meat they make, and the gas that they produce. These researchers are also interested in manure: they have been spotted, on occasion, "running around with buckets on the end of a stick with animals that they'd spray-painted giant numbers onto the side of, trying to collect their dung and urine, so that they could track the fate of nutrients."

The hills roll downward from the manor house until they go to other farms. They ultimately give way to Dartmoor, the gloomy highland. Dartmoor is a place where the wind whips mountains and the devil, people say, once paid a visit. It inspired Agatha Christie; it is where Emily Trefusis tracked the death of Captain Trevelyan, where

Sherlock Holmes once went to find "a foul thing, a great, black beast, shaped like a hound, yet larger than any hound that ever mortal eye has rested upon." On clear days, a modern Holmes could peer down from one of Dartmoor's pinnacles and see, at North Wyke, researchers engaging invisible worlds—probing microscopic life, capturing silent air, testing hidden minerals. These researchers have recorded more than 80 million measurements since 2010, but their work began long before the twenty-first century. North Wyke is a part of Rothamsted Research, the institution founded by John Bennet Lawes, the superphosphate pioneer. After he made superphosphate in 1842, Lawes studied fertilizers for fifty-eight more years. The early sale of phosphate fertilizers funds the research that Rothamsted conducts today, and in 2009, the institute integrated North Wyke into its organization. The work that Rothamsted's researchers are doing at North Wyke has been going on for nearly two hundred years, but their focus has changed considerably. Today their research is designed, in its totality, to answer a single, pressing question: How can humans reenter the cycles of the earth?

The modern history of loss in Nauru conceals a deeper narrative of hope. At least for centuries, and probably for millennia, people on Nauru crafted social structures that preserved the island and themselves. Their land was tiny and limited. The sea separated Nauruans from other people. But on a languid rock far out in the ocean, people not only survived—they thrived within the limits of the earth.

They were not alone. The fertilizing practices of the ancient agricultural societies—around the Mediterranean and throughout the Middle East, at the base of the Himalayas and in the basin of the Amazon—were designed to ensure survival within the parameters of the planet. Traditions that had developed over millennia were undone in decades whenever colonial or imperial conquerors arrived on the scene. When much of South America fell to the Spanish Empire, people were forced into two extractive industries: mining rare minerals and raising cash crops on large plantations. Simón Bolívar's political philosophy was born, in part, of outrage over this transformation. In 1815, Bolívar lamented that the indigenous people of South America were reduced to raising vast herds of cattle, mining gold, and growing indigo, grain, coffee, sugarcane, cacao, and cotton for their colonial rulers. "Is it not an outrage and a violation of human rights to expect a land so splendidly endowed, so vast, rich, and populous, to remain merely passive?" he asked. Remedying these injustices and others, Bolívar's revolutions were intended to restore the land to the people.

In Mexico, hundreds of varieties of corn once sustained legions of small farmers who produced unique crops of high culinary value and nutritional worth. In the centuries before the green revolution—when the Rockefeller Foundation and the U.S. government reshaped the agriculture of the world—corn was a food that grew in ears two feet long and purple, six inches and red, multicolored, squat, small-kerneled, large-kerneled, blue, yellow, bitter, sweet, creamy, and coarse. The decline of Mexico's corn culture came as a result of the green revolution and also of the North American Free Trade Agreement, which encouraged the flow of cheap corn imports from the United States, undercutting these small growers. Diverse varieties of Mexican corn were replaced by U.S. monocultures that tasted worse, contained fewer nutrients, and provided profits to businessmen rather than to farmers. The green revolution standardized a single kind of corn that responded well to fertilizer, was yellow and about a foot long, and contained ample carbohydrates but little other nutrition. Millions of Mexican farmers, forced off their land by the economics of free trade, fled to the United States and turned to low-wage labor, often working for the same corporations that had deprived them of their livelihoods back home. The new U.S.-bred corn grew less efficiently than the corn that had been raised in the old, carefully balanced systems. A corporate field planted with corn today must be, on average, three-quarters larger than a diversely planted field in order to produce a comparable amount of food.

In Australia, where the land was destroyed by colonialism and rebuilt with imperialism, agriculture has become a source of pride and an important cultural association. But Australia's agricultural renaissance, fed by Nauru, was based largely on a misconception. There is a flawed notion that farming was not possible on the landmass until rock phosphate was imported. The commonly accepted idea that phosphate transformed a barren wasteland into a prosperous paradise is simply not true. When the British began to colonize Australia in 1788, they interrupted a long tradition of indigenous agriculture that had carefully protected the vulnerable soils of the continent. Aboriginal people inhabiting the landmass for millennia nurtured the ground, ensuring a supply of food from one generation to the next. The indigenous people of Australia traded seeds. They applied natural fertilizers to the soil. They grew grains to make bread, raised yams in fertile riverbeds, and cleared fields for cultivation. In the desert, they grew nardoo ferns, which survive in empty lake beds and other dry conditions. They grew rice. They engaged in fish farming. They tracked and trapped animals, including kangaroos and emus, using techniques that blurred the line between hunting and husbandry. These animals were suited to their conditions; kangaroos, for example, hopped lightly without compacting the ground.

The soil of Australia is immensely old—older than the soils of Europe, Africa, or the Americas. Even under Aboriginal cultivation, it suffered deficiencies of many nutrients, including phosphate, which existed only at very low levels. But, in order to survive, Aboriginal Australians learned to work within these conditions, safeguarding the phosphate that was present. For millennia, they regularly burned portions of the landscape, concentrating phosphorus below the ground while limiting the risk of uncontrollable wildfires above it. They protected the soil by mounding it, growing gardens that developed over time. Within those gardens, they applied ash and compost, which aerated the soil and expanded its supply of nutrients. The most fertile lands were kept clear and healthy, while less fertile lands were planted with trees. To prevent erosion, the land was hoed across the curve of the slope. Water was conserved through the construction of dams. Eighteenth-century European colonizers found soil that was easily cultivated. It had been irrigated. "The unusual quality and friability of the soil was reported by many colonists in the first years of settlement," the historian Bruce Pascoe, who descended from Aboriginal ancestry, wrote in *Dark Emu,* his 2014 book, which compiled these pieces of evidence of precolonial agriculture on the continent, drawing controversy in the process. The genocide carried out by white colonizers against Australian Aboriginal people included the repetitive consumption and destruction of the land through farming practices that depleted the soil of its nutrients, particularly phosphate. Local plants were killed. The settlers brought with them new animals that destroyed the ground and new plants that were not adapted to survive. The soils hardened. Flooding worsened, and nutrients flowed away. The long-protected soils disappeared. Despite millennia of Aboriginal use, one observer noted that pastures often failed after three years of colonial cultivation.

The result of this failure was a myth that Australia was inhospitable to agriculture. It was dry, desolate, and barren, it was devoid of phosphate, and Nauru's exports made it grow. In reality, Australia needed Nauru only because local knowledge had been replaced by European ignorance. As time passed and airplanes came to spray phosphate fertilizer onto farm fields in Australia and New Zealand, the soil in those places weakened. Chemicals began to accumulate, soil eroded, and floods increased in frequency. Dust storms in Australia crossed the Tasman Sea, creating burnt-red mountains in New Zealand. Oceania became a European-style agriculture powerhouse, but its farmers paid for their ambitions with borrowed time.

Throughout the twentieth century, Nauruan phosphate was exported to more places than Australia and New Zealand. Ships carried it to Japan, to Europe, and to North America, and some ships took the rock to England, where a portion of it reached the premises of the Lawes Chemical Manure Company, the firm that had been founded by John Bennet Lawes in 1842. In a factory at Deptford, on a bend in the Thames, Lawes turned phosphate rock into phosphate fertilizer. In 1872, he sold his company for £300,000. He used the proceeds of the sale to endow his research at Rothamsted with funds to last in perpetuity. Today at Rothamsted Research, hundreds of scientists are continuing Lawes's work. They have transformed his country mansion into a conference hall, his manor into an internationally regarded scientific institution. They are running experiments with ages verging on two centuries, explorations of the evolution of soil over time. "At Rothamsted, we're very proud of our heritage, of John Bennet Lawes. And I always tell everyone, it's down to phosphorus fertilizer, the reason we've all got jobs today. Because that's what he founded the research institute on—that's where he made his money," Martin Blackwell, a soil biogeochemist at Rothamsted, told me in 2020. "We still don't understand phosphorus properly. We've been looking at that for a hundred and seventy-eight years now. . . . There's a whole load of work that needs to be done."

The legacy of Lawes was alive in the advances of the twentieth century. The green revolution, guided in its early years by the Rockefeller Foundation and the U.S. government, had promised great prosperity to the world—but the harvest of 1968, striking in its size, was only a temporary event. Historic wheat harvests arrived in India, yields surged elsewhere, and William Gaud gave the green revolution its commonly applied name, but as soon as the harvest ended, tensions began to grow in the countries where production had been increased. The irrigation, the fertilizers, and the farm equipment brought to increase yields had typically been sent to larger farms rather than to smaller ones, and the disparity worsened rural inequality. An influx of cheap grain onto the market scrambled geopolitical realities and undercut poor farmers. The immediate effect, in many communities, was that smaller growers struggled, accumulated debt, and ultimately lost their land to wealthier investors. Skilled labor was replaced by machines, crops were homogenized, and local knowledge was lost. Rural workers fled to the cities, where many failed to find jobs. Millions of people lost their livelihoods. Unrest grew worldwide.

One by one, the green revolution governments either toppled or turned authoritarian. Ayub Khan, Pakistan's president, who had promoted agricultural technology, fell out of favor and resigned in 1969. Dudley Senanayake, the prime minister of Ceylon, saw his majority disappear in 1970, as high expectations set by the green revolution were wiped out by storms. The president of the Philippines—the headquarters of the International Rice Research Institute—declared martial law in 1972, while Indira Gandhi, the Indian prime minister who had trumpeted the arrival of chemical fertilizers, assumed emergency powers in 1975, ultimately imprisoning 36,000 people. (Nine years later, she was murdered by her own guards.) Military juntas in Thailand and Indonesia took control of fertilizer and trade. "Throughout South and Southeast Asia, the green revolution would be linked in memory with dictatorship and military corruption," the historian Nick Cullather wrote in a history of the period. The illusion of progress had lasted only a year.

The land itself was degraded by the influx of unnatural techniques and processes. Heavy irrigation caused salt to accumulate above the ground. Pesticides killed organisms in the soil. Hulking machines, brought in to manage the land, destroyed it. They caused soils to compact, air pockets to disappear, and microorganisms to die as a result. New seed varieties were not well suited to the varied conditions of the land. They were vulnerable to changing weather conditions, and as a result, harvests were unstable from year to year. Crops were planted closer together, at higher frequency and with far less genetic diversity than before, creating an opportunity for insects and diseases to spread. Researchers tried to stave off destruction by isolating a gene that would defend against the pests, but the frequent migration of the insects and the constant mutation of disease meant that each batch of seeds quickly lost its effectiveness. The process had to be repeated annually.

Outside of the farms themselves, a dramatic symptom of fertilizer overuse soon arrived in public view. Excess fertilizers were draining from the fields, entering rivers, and building up in lakes and ponds downstream. The fertilizers spurred the growth of algae, which occupied the water and then died. The bacteria that decomposed the algae consumed the oxygen in the water, leaving behind murky, bacterial pools where fish and other creatures could not live. Eutrophication, the scientific term for this event, results in hypoxia, a state in which an aquatic system lacks oxygen. People took to calling these places "dead zones." In 1968, scientists showed that the scourge of eutrophication was connected to the use of superphosphate. The biogeochemists Jack Vallentyne and David Schindler chose a quiet lake in northwestern Ontario, divided it in half, and added carbon and nitrates to the whole lake but phosphate to just one side. Within months, the results were clear: phosphate caused eutrophication. The half of the lake with added phosphate was covered in a thick algal bloom, while the other half remained clear. A photo of the lake, an iconic image of the environmental movement, showed with clarity that inland dead zones were a product of phosphate fertilizer.

The decline of yields, the rise of dictators, and the decimation of lakes all combined to shift the scientific view of phosphate. At Rothamsted, scientists grappled with the force their predecessors had unleashed. "From my own perspective," Martin Blackwell told me, this realization came from two places. "One is, it didn't always work. The recommendations were developed, and they worked for a while. But then you start degrading your soils, and you have to keep putting more and more fertilizer on to get those yields, because you're depleting it. . . . And then, I think it was really in the seventies or eighties, we saw these massive environmental impacts. We suddenly became aware of the wider damage caused by putting too much nitrogen and phosphorus into our soils." In 2021, when I visited North Wyke, the soil microbiologist Andy Neal was blunt about what happened. "We've trashed the planet. Now it's our responsibility to try to make things better," he said.

A discussion of phosphorus is a discussion of sewage, of agriculture, and of ways of life. In two centuries, our use of sewage has changed. Our mode of farming has shifted. In 2009, North Wyke became a part of Rothamsted Research. Whereas Rothamsted had once focused on the growth of plants through geology, its scientists now wanted to rebuild connections between fertilizer and life. In the west of England, where North Wyke is located, livestock predominate. Cows and sheep graze the land in lieu of many crops. The scientists divided North Wyke's fields into experimental plots, and they measured everything. Eddy covariance towers were constructed to measure the emissions of greenhouse gases from fields with grazing animals, and LI-COR chambers began to measure emissions from particular patches of ground. "We look at three aspects of sustainability: economic, social, and environmental," Melanie Wright, whose work at North Wyke involves making connections between farmers and researchers, told me when I visited. "Then we also look at the social sustainability of them. And generally, for that, we're looking at the actual quality of the product they produce. Is it good for mankind? Does it produce a product which is high in the essential nutrients that we need as part of a healthy, balanced diet?"

The overarching hypothesis of North Wyke's research is that "it is possible to produce nutritious food in a sustainable manner—as evidenced by economic, environmental and social metrics we define and measure—even under a growing population and a changing climate." Accomplishing that goal requires restoring the phosphorus cycle, and restoring the phosphorus cycle requires changing the way that agriculture works today. Data, in the twentieth century, made farming more mechanical, but scientists at North Wyke believe that data can take it back again. The central studied area of North Wyke is called the North Wyke Farm Platform, and it totals 163 acres of land. It is the most heavily instrumented beef and sheep farm in the world. Since 2010, the soil scientists, atmospheric chemists, livestock scientists, hydrologists, and data scientists working there have demonstrated how soil microbes create pore space in aggregates, allowing nutrients to move more freely. They have defined relationships between animals and the ground. They use a Haldrup plot combine to take grass samples. They use a forage harvester when they want to weigh the grass. There are five and a half miles of French drains constructed along the lower boundaries of each of the Farm Platform's five catchments; that piping captures runoff from the fields and channels it into fifteen laboratories. Phosphorus—deposited by animals, absorbed by the roots of plants, held within the soil, washed off, sometimes, by the rain—is measured by this system. The phosphorus cycle on the Farm Platform has been quantified.

If a single key conclusion from the phosphorus tests at North Wyke could be drawn, it might be this: the flow of phosphorus through soil is complex. When phosphate fertilizer is applied, most of it is either held by soil particles or taken up by microbes. Over time, these microbes release the phosphorus that they accumulated, and they also help to make available other bits of phosphorus that bonded to the soil. Take away those organisms, and the nutrient cycle suffers. On the whole, crops generally absorb one-fifth of the phosphate that is applied to the soil. The rest is fixed, rendered insoluble, and held within the ground. It is only available over decades or centuries, and then only through the efforts of bacteria and fungi in the soil. During the green revolution, those microorganisms were not nurtured; phosphorus accumulated while remaining inaccessible. Around the world today, 70 percent of cropland contains more phosphate than it needs—although many of those soils, depleted of life, do not provide their phosphate to the crops that grow within them.

There are other factors. Phosphorus is not a very mobile nutrient, which means it does not travel on its own. It should be placed as close as possible to roots: scattering phosphate on top of the soil—which is known as broadcasting—is not a reliable practice. Phosphorus will be more easily absorbed if the soil has just the right acidity, as well, making lime treatment sometimes necessary. (If the soil is too acidic, or if it is not acidic enough, phosphorus will bind to other elements, leaving it, as chemists tend to say of such tightly bonded molecules, unavailable.) Too much phosphate in the soil can harm the development of plants and can limit the uptake of other nutrients. Phosphate passes through the soil in a way that resembles an ecosystem rather than a factory. The question is not how to add enough phosphate to farmland, but how to use what is already there. "I think it would be fair to say that in the past, the production of phosphate fertilizers has fed the planet, but now we've saturated our environment with it, and we need to find ways of making it more circular, of recycling it," Martin Blackwell, who, along with Andy Neal, is stationed at North Wyke, told me. "We've put enough phosphorus into the environment. We can do a pretty good job at reducing the amount of rock phosphate–derived fertilizer that we need."

When Lawes was making superphosphate in 1842, he was aware of the potential for unintended consequences. As he tested fertilizer on his estate, he was working with a professional scientist, Sir Henry Gilbert, to create a broader set of experiments. They wanted to know the long-term effects of superphosphate. Between 1843 and

1856, the two men fashioned eight experiments that were designed to run continuously on plots of planted land. Across the Rothamsted estate, the two men planted wheat in 1844, barley in 1852, and hay in 1856. On various other plots they planted grass, root crops, and clover. The practices varied, but the plan was simple: the same procedures would be carried out on all the plots in perpetuity, and the health and output of the land would be continuously measured. Of the eight experiments Lawes and Gilbert started, six are still in progress. They constitute the longest-running agricultural experiments in the world.

Several decades ago, after a century of treatment, Rothamsted's researchers found a straightforward answer to the question of fertilizer's long-term effects. In the Broadbalk, Hoosfield, and Barnfield experiments, wheat, barley, and turnips were grown in plots of land that were treated annually with varying levels of manure and synthetic fertilizers. Plants grew similarly when treated with organic and inorganic fertilizers, but the soil in which they grew changed significantly over time. In the plots that had been treated with manure, compared with those that had received synthetic treatments since the nineteenth century, twice as much organic matter was present. Manure had been recycled, no inputs had been purchased, and the soil was healthy after a century of providing food. Organic matter, it turned out, was actually an important part of the process. It had fed the microbes, which not only transferred phosphorus to plants but also staved off disease, regulated the balance of chemicals in the soil, and provided a diverse array of nutrients. It had retained water and maintained the structure of the soil. Just as fiber feeds microbes in the human gut—those microbes weigh, on average, 4.5 pounds—fibrous fertilizers had fed the microbes in the soil. And just as microbes in the human body help us fight disease, so do microbes in the soil protect the plants that grow there.

Following these discoveries, Rothamsted's researchers created the Woburn Organic Manuring Experiment, a new long-term study. Beginning in 1964, crops were grown with more varieties of either organic or synthetic fertilizers. Over decades of growth and measurement, the amount of organic matter in the soils of the synthetically fertilized plots steadily dropped, while it rose in the other plots. "The whole fertilizer issue is that sometime in the last one hundred years, we forgot that the soil requires a regular throughput of organic material. We focused on feeding the plant and increasing yield," Andy Neal, the soil microbiologist, said. "The knowledge of how soil works was not there when these guys were trying to figure out how to feed people. John Bennet Lawes and Norman Borlaug didn't understand the complexity of the soil."

The soil functions as a living organism that preserves the world of a billion years ago while sustaining lives that will continue far into the future. The fungi and bacteria that populate the ground constitute an astonishing set of living beings. Fungi, which can grow in acid, liquefy wood, and drill into rocks with a force far greater than that of any human, once made up the dominant life-forms on earth: giant fungi pillars rose during the Devonian, eclipsing other beings. In modern times, fungi have been found living two miles below the ground and in other extreme locations. They are responsible for the production of citric acid, for the creation of many cheeses, and for the production of beer and detergents and soy sauce and chocolate and bread. Of the twenty most-used pharmaceuticals in the world today, half were derived from fungi. Bacteria, the partners of fungi in the soil, are similarly grand actors. Bacteria evolve by replication. Unlike humans, who change genes slowly through reproduction, bacteria freely exchange their DNA with one another, continually building newer and better versions of themselves. Because they are particularly good at evolution, they helped to fashion modern life. Mitochondria, which generate energy in our cells, evolved from bacteria that once swam free. Floating inside every cell, bacteria preserve the past. They also design the future: in four days, given unlimited resources, a single bacterium would produce more offspring than there are protons in the universe.

A spoonful of healthy soil contains a billion bacteria, which work to transfer phosphorus to the roots of plants. In the soil, phosphorus and other nutrients are locked inside crystals collected as particles of sand. That sand, long ago, weathered from mountains, making its way to the soil, but it remained inert. Fungi break sand slowly, releasing the ingredients each grain holds. Bacteria consume those nutrients and hold them in the soil when they die. Mycorrhizal fungi form nets that extend for miles, connecting the roots of plants, creating a vast community. These fungi also process phosphate, transferring it out of the soil and into the roots they bind together. Fungi exist, in nature, at the whim of plants: when more organisms are needed, more fungi will appear. When fewer are needed, fewer exist. Without fungi and bacteria, less phosphorus is removed from sand. While some weathering of sand still occurs through chemical and physical means, the rest of the phosphorus remains trapped. It does not cycle through the soil. It does not reach the roots of plants. The erosion of the mountains reaches a dead end.

The green revolution enjoys a global reputation for saving millions from starvation, protecting human health, and maintaining political stability in the process. Norman Borlaug was awarded the Nobel Peace Prize for his leadership in the effort, and the U.S. government continues to support the same industrial style of agriculture at home and around the world. It is hard to argue with the logic that greater efficiency feeds more people—but besides its environmental effects, the green revolution has also left a complicated nutritional legacy. While it likely helped to increase the total caloric capacity of agriculture around the world, the green revolution is also generally credited with reducing the nutrient content of grains. Fertilizers were added, but soils need more than conventional mixtures of nitrogen, phosphorus, and potassium: in order for plants to grow, they also need calcium, sulfur, magnesium, boron, chlorine, copper, iron, manganese, molybdenum, and zinc, among others. In a healthy soil, fungi and bacteria obtain these elements from sand. In an unhealthy soil, they do not.

The loss of life in the soil has limited a wide range of nutrients that used to be converted into food. Fertilizers have remedied phosphorus, nitrogen, and potassium deficiencies, growing more cycles of crops over shorter periods of time and increasing yields substantially. That intensive cultivation has led to a loss of other necessary elements. Today, many soils across Asia lack zinc. In India, soil scientists estimate that half of all farmland is deficient in zinc and one-third is deficient in boron. Plants that lack boron often stunt; plants that lack zinc turn yellow and slowly lose their health. A human diet that relies on plants grown in soil that lacks these nutrients will result in a weakened skeletal system, which requires boron to lay down new tissue, and immune system, which relies on zinc. Eventually, an acutely zinc-deficient person might experience nausea, diarrhea, loss of appetite, greater vulnerability to illness, and stunted growth. A three-year greenhouse study of depleted soils in Zimbabwe showed that application of manures—which contain a greater diversity of nutrient-containing microorganisms—corrected zinc deficiency in soils and also alleviated a lack of calcium in the sandy, overused strata. When plants do not consume enough calcium, their leaves turn brown and become brittle. When humans do not consume enough calcium, our skin weakens and our bones break.

In Australia and New Zealand, in much of sub-Saharan Africa, and in parts of the Amazon basin, soil sulfur deficiency is a common problem. A plant that lacks sulfur will lose its color. The tips of its youngest leaves will begin to die, and the plant will eventually cease growing. Sulfur is crucial for metabolizing amino acids in humans, and a diet low in sulfur affects the body in myriad ways, including by harming the skin, joints, and brain.

Magnesium, which is necessary for the maintenance of the human skeletal system, supports the growth of leaves in plants by enabling photosynthesis. In recent decades, the magnesium content of grains such as wheat and corn has decreased markedly around the globe. Similar declines are evident in iron and zinc, and they are attributed to the spread of industrialized agriculture. Today, most Americans do not consume enough magnesium; this trend is also seen in other wealthy countries. Magnesium deficiency in humans is linked to cardiovascular disease, because magnesium ions keep the heart relaxed and beating at a regular rate. The decline of magnesium in the American diet tracks closely with the rise of cardiovascular disease across the country. In the 1960s, when dietary magnesium uptake reached its lowest level, deaths from cardiovascular disease peaked.

The success of early agriculture rested on the accumulation of nutrients in order to produce concentrated amounts of food. Just like today, infertile soils could not produce crops—and fertile soils, unreplenished, soon began to fail. The diffuse products of nature had to be collected and condensed within individual plots of land. Crops were chosen to grow in these conditions and satisfy human hunger. Slowly at first—and quickly later—food diversity was lost. Vitamins, minerals, and micronutrients disappeared from crop plants—gone were lycopene, sulforaphane, silymarin, cynarin, inulin, ellagic acid, hesperidin, flavanone, naringenin, hesperetin, naringin, beta-carotene, glucosinolates, isoflavones, proanthocyanidins, and carotenoids, nutrients that plants had developed to protect themselves. "Over the course of ten thousand years of agriculture, our farming ancestors managed to remove the bitterness from most of our greens," the food writer Jo Robinson noted in her 2013 book *Eating on the Wild Side,* a comprehensive account. "Unwittingly, though, when they removed the bitterness, they were also stripping away a host of highly beneficial phytonutrients that happen to have a bitter, astringent, or sour taste. Our mild-to-a-fault iceberg lettuce, for example, has one-fortieth as many bionutrients as bitter dandelion greens. Calcium is bitter as well, so the calcium content of our modern greens is also relatively low. This could be one of the reasons that osteoporosis now afflicts so many older Americans. In 2011, forty-four million individuals were diagnosed with low bone density or with osteoporosis, placing them at high risk for fractures. Hunter-gatherers who consumed calcium-rich wild greens had much denser bones than we do today, despite the fact that they consumed no dairy products."

The green revolution only magnified a trend that had long since taken hold. Modern food and medicine have divided into parts what used to be the whole of the human diet. People once ate widely, picking nuts, roots, fruits, leaves, seeds, and bark from hundreds, sometimes thousands, of plants. The organisms they ate grew in different variations of soil composition and sunlight, absorbing a wide range of beneficial nutrients that were passed upward through the food chain. Animals that ate those plants and people who ate those animals benefited from the diversity of the soil. Just as modern medicines are made from plants, fungi, and other natural organisms—including the majority of cancer drugs and the vast majority of popular prescription medications—those organisms, when eaten as part of a regular diet, supported human health. Plants can get sunburned, so they have learned to produce chemicals that protect from ultraviolet radiation. Trees can get cancer, so they produce chemicals to protect from dangerous growth. The diets of hunter-gatherers consisted of wild precursors to today's foods. People ate different foods each month as the cycling of seasons provided new varieties of plants. They absorbed a bounty of nutrients.

Agriculture limited plant diversity, but societies still augmented local foods with varied wild options—picking berries in the spring, searching for truffles, and hunting game at various times of year. Only after the nineteenth-century fertilizer revolution, when industrialization pulled people from the countryside and farms became more mechanized, did those cultural institutions begin to fade away. As farms became bigger and more globalized, they became less diverse, and people began to rely on them for all of their food, not just some of it. The twentieth century brought seed standardization. New types of crops were developed that maximized fast growth and sweetness and resistance to pesticides. The same varieties were fed to everybody. The health and diversity of the soils declined, limiting the nutrients that would be eaten. Animals, instead of grazing in pastures and forests, were raised indoors, in the dark, where they lived on antibiotics and nutritionally bereft grains.

Compared to diets that rely on a diversity of plant and animal life—including those of many indigenous communities—the standard American diet relies on only several species of plants. Research suggests that hunter-gatherers barely suffered from the chronic diseases that now kill most Americans. Many lived to reach old age. Infant mortality was high until recently, infectious diseases were poorly treated, and accidents caused numerous deaths, but for thousands of years, a person who reached adulthood could reasonably expect to live for several decades after that. Many people grew old, but few people suffered from the conditions that afflict the aging classes of wealthy societies today. Members of modern indigenous communities that have not been disrupted by industrial agriculture generally experience a sharply lowered risk of scurvy, pellagra, beriberi, goiter, anemia, type 2 diabetes, coronary heart disease, hypertension, osteoporosis, asthma, liver disease, gout, myopia, cavities, and hearing loss. Meanwhile, many of those diseases, including diabetes and heart disease, have become the main causes of death in the United States. Globally, heart disease accounts for nearly one out of six deaths every year.

Industrialization was not a foreordained conclusion of human agriculture. When Henslow picked up phosphate at the Bawdsey Cliff, there were other options. In the countryside of England, the privatization of public land had led to the instability that Henslow sought to end. One response to this privatization was the provision of allotments, small plots of land within villages that could be rented by individuals and used to grow food. In East Anglia between 1830 and 1851, more than fifty parishes began new allotment programs. In 1845, Henslow himself proposed bringing the allotment system to Hitcham—in order, he wrote, "to palliate the evils under which our poor are labouring." The farmers of the parish, who employed the laborers for whom the allotments were intended, opposed the plan out of fear that such reform would reduce the people's need to work for wages. "I am sorry to say it appears that you are one of those Philanthropical Gentlemen who wish to make themselves popular with the lower class of society at the expense of the Farmer," one angry farmer wrote to Henslow in response to the proposed allotment program. "I will never cart another ton of coals for the poor so long as this sub-letting system is in existence."

But Henslow beat the farmers. By 1849, he had procured fifty-two plots of land, a quarter of an acre each, which were made available to a lengthy waiting list of people who wanted to grow their own food. "I think he cared somewhat less about science, and more for his parishioners," Jenyns wrote of these initiatives. "When speaking of his allotments, his parish children, and plans of amusing and instructing them, he would always kindle up with interest and enjoyment. I remember one trifling fact which seemed to me highly characteristic of the man: in one of the bad years for the potato, I asked him how his crop had fared; but after a little talk I perceived that, in fact, he knew nothing about his own potatoes, but seemed to know exactly what sort of crop there was in the garden of almost every poor man in his parish."

The allotment system was a social solution to a social problem, while superphosphate was a technological solution to it. Landowners opposed the allotment system because they feared a world in which such social problems had been solved. They supported the use of superphosphate because its cost kept power in the hands of those who already held it. Henslow secured his allotments, but the wealthy won the broader argument: superphosphate became a new standard of agricultural practice.

A century later, the process repeated itself. The green revolution was named in contrast to the so-called red revolutions, communist uprisings that were sweeping through the world in the twentieth century. In the 1930s, Lázaro Cárdenas, the president of Mexico, announced plans to redistribute land from the wealthy to the poor, and conservatives asked the U.S. government for help. The architects of the green revolution argued that the welfare of the poor would be improved by technology rather than by reforms of landownership, and Cárdenas's plans were never carried out. Repeatedly, policymakers chose to respond to social problems with technological solutions. In 2006, it happened again.

The Rockefeller Foundation funded the first green revolution and, in 2006, it joined with the Gates Foundation to create a new one: the Alliance for a New Green Revolution in Africa, with the support of the United Nations, set out to turn subsistence farmers into businessmen. "The original Green Revolution was a huge success in many parts of the world," the president of the Rockefeller Foundation said at the announcement. "Working with the Bill & Melinda Gates Foundation and with African leaders, farmers and scientists, we're committed to launching an African Green Revolution." "These strategies have the potential to transform the lives and health of millions of families," Bill Gates added. The goal of AGRA, for the thirteen countries where it worked, was to double crop productivity and halve food insecurity by 2020. It would do so mainly through the introduction of chemical fertilizers and new genetic technology, doing in Africa what had been done in Asia fifty years before.

Like the first green revolution, this new effort relied on a slate of policies related to technological change. Marshaling its clout across the continent, the organization used an annual $50 million investment to push national governments to spend about a billion dollars each year on AGRA's priorities. This money generally benefited larger farms that, run by well-off men, were not representative of the agricultural landscape. (Although roughly half of all farmers around the world are women, 85 percent of landholders are male. It is harder for farmers to invest in the long-term health of soil when their right to use the land remains uncertain.) One early success of the campaign to promote industrial agriculture was the Abuja Declaration on Fertilizer for the African Green Revolution, adopted by African heads of state in 2006, which aimed to increase fertilizer use sixfold within nine years. In order to achieve the goal of increased fertilizer usage and greater mechanization of farming overall, government funding came with certain stipulations. Depending on the project, farmers might be required to stop saving seeds and to buy genetically modified seeds instead, or they might be required to purchase synthetic fertilizers, or to end the practice of intercropping. With AGRA setting the agenda and the national governments enforcing these new rules, agriculture on the continent was set to change.

AGRA's approach mirrored that of its predecessor, but in the decades that had passed, the world had evolved. Due to the work of scientists at Rothamsted and other institutions, skepticism of mechanical agriculture had accrued. Racist ideologies, torn apart in the postcolonial era, gave way to greater understanding of local agricultural knowledge. In response to the ravages of the green revolution and the excesses of corporate agriculture, resistance to agribusiness had strengthened. The United Nations, which had cheered on the green revolution—and, initially, its Gates-led successor—began to express more doubt. As the new millennium wore on, the

U.N. Food and Agriculture Organization released reports that touted the benefits of more ecological farming, while the U.N. Intergovernmental Panel on Climate Change increasingly turned its focus to the negative global effects of industrial agriculture. In 2016, the International Panel of Experts on Sustainable Food Systems, which had been founded by Olivier De Schutter, the former U.N. special rapporteur on the right to food, released a report that endorsed a widespread transition away from global monocultures. By then, even Bill Gates had acknowledged the concerns of his program's skeptics. At the 2009 World Food Prize Symposium—held a month and three days after the death of Norman Borlaug—he spoke about his new goals for Africa. "Many environmental voices have highlighted that there were excesses in the original Green Revolution," he said. "These are important points."

The AGRA effort, however, continued to favor outside technologies over local knowledge. It approached agriculture the way the green revolution had before, and it faced similar criticism. Yields were changing slowly, and positive outcomes were rare. In 2018, an international group of agricultural organizations analyzed the changes in both productivity and extreme poverty in the thirteen participant countries for twelve years before and twelve years after AGRA was introduced, finding them unchanged. In nine of those countries, hunger had worsened rather than improved. "Only one country, Ethiopia, shows anything resembling the kind of yield growth and hunger reduction Green Revolution proponents promised, with a 73 percent increase in productivity and a 29 percent decrease in the number of those going hungry. Still, neither is on track to meet AGRA's goal of doubling productivity or halving the number of hungry" people, the report noted.

As difficulties mounted, AGRA stopped providing data on its progress, though it insisted that its goals would be met by 2020, a deadline that then passed without an update. Some of AGRA's donors requested an evaluation of the program, and in February 2022—two years after the organization's self-imposed deadline on the continent—the evaluation confirmed criticisms of the organization's effectiveness. Overall, the evaluation found, AGRA had not achieved its goals of improving farmer outcomes in the countries where it operated. "This finding is an expected outcome and a true reflection of the realities that farmers, AGRA, and other institutions that support farmers today live with daily," AGRA wrote in response. "We must therefore rethink our models and focus our support, and that of our partners, on building resilience and adaptation specifically for smallholder farmers."

Farmers, by then, had had enough. A month after the donor report was released, the Southern African Faith Communities' Environment Institute, an influential group, wrote to AGRA's funders, including the Gates Foundation and the U.S. Agency for International Development, urging an end to the program. That September, a group of organizations representing farmers and faith communities across Africa released a statement noting that AGRA had "pushed a development model that reinforces dependency on foreign inputs, such as expensive fertilizer, undermining the resilience of African food systems." One week later, at AGRA's annual meeting, the Africa Green Revolution Forum renamed itself Africa's Food Systems Forum. AGRA rebranded itself too, opting to go only by its acronym. The organization released a five-year plan for the years 2023 through 2028. Throughout the thirty-two-page document, the term "green revolution" was never used.

While the phosphorus cycle on farms is more than a system of inputs and outputs, it is possible to analyze what is put in and what comes out. About 17.5 megatonnes of phosphorus—19 million U.S. tons—are mined each year, of which about 15 are made into fertilizers. (This is a measurement of phosphorus, not phosphate: by weight, a phosphate molecule is roughly one-third phosphorus and two-thirds oxygen.) Humans excrete 3 megatonnes of phosphorus each year, of which only 10 percent is then recycled. Farm animals excrete 15 megatonnes of phosphorus; out of that, 7 megatonnes go to waste. One megatonne a year is lost to food waste. Eight are lost through erosion off of farms. Each year, the amount of phosphorus that slips away through all these means—wasted sewage, unused manure, food waste, and erosion—totals more than the amount that is mined. If phosphorus were reused as it once was—and as it still is in some communities around the world—mined sources would be all but unnecessary. So why is global agriculture still reliant on phosphate rock?

The benefits of artificial fertilizer lie in the creation of profits rather than the production of food. Globally, two-thirds of phosphate fertilizers are used on cereal crops, and in wealthy nations three-fifths of all cereal crops go to feed animals. A growing proportion of the remainder is used to make biofuels. While large farms, fed with ample fertilizers, dominate this kind of production, they struggle when it comes to growing food. It is small farms, not large ones, that produce food more efficiently. Collectively, around the world, agribusiness revenue totals $3 trillion a year. The owners, operators, and investors of these companies—and the politicians who profit from their largesse—benefit from this status quo.

The governments of the world devote significant funds to subsidizing industrial agriculture. During the latter part of the twentieth century, world leaders crafted a system of subsidized inputs designed to complete an industrial farming transformation. Those policies are still in place. The Chinese government's fertilizer subsidy amounts to roughly $19 billion annually. In India, fertilizer prices are kept artificially low, and the government pays the difference. In 2021, the price of phosphate fertilizer—in the form of diammonium phosphate—was kept at a 55 percent discount nationwide. The nitrogen fertilizer urea, whose price is strictly capped, retailed at a 92 percent discount. That year, the Indian government's fertilizer subsidy surpassed $20 billion. The global fertilizer market in 2021 totaled $193 billion. The governments of India and China alone—the governments, not the countries as a whole—contributed 20 percent.

That 20 percent does not account for the United States, where large subsidies for monocultures, particularly corn, support industrial production. Nor does it include Indonesia, where fertilizer subsidies amount to roughly $2 billion and take up half the country's agricultural budget. There, synthetic fertilizers are supposed to support small-scale farmers who grow rice to feed their communities—but a government report obtained by Reuters in 2016 showed that up to one-third of all subsidized fertilizers were misused, many of them going to palm oil and rubber plantations. Between 2011 and 2016, when fertilizer subsidies increased by 60 percent, the country's rice crop remained the same. Worldwide, the fertilizer industry is supported by state subsidies that create a false image of efficiency.

There is, perhaps counterintuitively, an inverse relationship between the size of a farm and the efficiency with which it operates. Compared with equivalent larger farms, small farms produce more food, suffer less damage from droughts and hurricanes, and provide greater nutritional variety to consumers. Rather than planting a single crop across a field, farmers create ecosystems containing different species of plants and different varieties within those species. They grow these plants on soils that they understand, providing their output to communities that they value. Their communities value them: about half of all food eaten by humans is grown by small-scale farmers. Even though small and large farms produce food with relative parity overall, larger farms occupy more land, control more capital, and consume the vast majority of pesticides, herbicides, and synthetic fertilizers used worldwide. But in some places, this is changing.

Putting waste at the center of farming—where it always, until recently, had been—is a democratic action. Waste is universally emitted and individually controlled. Some governments have recognized this. In the state of Andhra Pradesh, in India, where the ground was devastated by the green revolution, the government has reversed course in recent years, supporting farmers who foster connections with the land. The result has been higher yields, more food, and less poverty. The World Bank, working with the government of India, is trying to undo parts of the green revolution entirely, returning farmers to their prerevolution seeds. In Cuba, with the support of the organization Asociación Nacional de Agricultores Pequeños, half of all small farmers have transitioned toward more ecological farming. Around the world, there are city gardening projects, efforts to improve the nutrition of food, and policies designed to protect small farmers. Largely as a response to phosphate-polluted water systems—Lake Vansjø, Lake Ringsjön, Lough Neagh, the Baltic Sea—countries in northwest Europe have implemented measures to reduce erosion from farmlands and reuse manure, resulting in steady annual declines in phosphate runoff. The Netherlands has created a manure trading system that tracks the inputs and outputs of nutrients on farms and mandates recycling. Only half of all humans have access to sanitation systems, and the developmental work done to improve that sorry statistic—to make clean toilets more accessible—ought to take into consideration the failures of the last sanitary revolution. This is, in fact, what is happening. In parts of India and South Africa, in Niger and Sweden, newly built sanitation systems turn waste into fertilizer, rather than placing it in a one-way channel to the sea.

Global organizations are changing too. La Via Campesina—"the Peasant's Way"—was founded in 1993 and has grown to include farmers in eighty-one countries who advocate, collectively and individually, for policies that support the rural poor. The farmers are bound together by a shared commitment to resisting the growth of farming corporations, and in 2018 they achieved a landmark victory: the United Nations, long a supporter of industrial agricultural policies, adopted the U.N. Declaration on the Rights of Peasants and Other People Working in Rural Areas, a statement on the importance of small farmers to the world.

On an individual level, the simplest way to restore the phosphorus cycle is to start a compost pile. Food scraps, piled together with leaves, sticks, and other plant materials, will yield rich soils. Turned occasionally, the parts will come together. They will darken and crumble. The scraps will become soil. Contained within this soil will be micronutrients held by microorganisms that thrive in the presence of organic matter. This soil will grow upon itself from year to year, creating prosperity that does not end.

As old age comes, the soil gives retirement. It provides a gentle garden, lush vegetables, poetry, sustenance, shade. When we die, we leave the soil richer than we found it. This is not an input industry; it is not a process that transfigures overnight. It is a process of creation that relies upon the circularity of life, the slow decay of bodies, the procedure of death that comes after the final light of being.

. . .

At Rothamsted, researchers are demonstrating the importance of such processes. The scientists at North Wyke are looking for ways of providing phosphate to plants through natural processes—and not through the application of phosphate rock. Rothamsted's research—perhaps surprisingly, given the organization's history—has shown the importance of organic matter, of mixed systems, of the usage of farmyard manure. Its scientists are pushing for the reuse of bones as fertilizer, for the recycling of organic waste. They are trying to find a way to re-create the cycles on which ecosystems rest. ("Nature had several billion years' head start on us," Melanie Wright said.) Andy Neal told me, "When we're doing this science, this research, it can look like we're reinventing the wheel. But when I think of our role, I think that we're trying to understand why these things that people observe happen the way they do."

In Japan, half of all human sewage is still made into fertilizer. In Windhoek, the capital of Namibia, wastewater has been purified and supplied to the population on an industrial scale since 1968, a project that one-third of the city relies on for its drinking water. Phosphorus and nitrogen present in Windhoek's sewage are reused as fertilizers. In Yemen, large buildings historically included waterless toilets that separated waste between urine, which was evaporated, and feces, which was used as fuel. After flush toilets were introduced, water shortages worsened. Through anthropological work, GPS mapping, and analysis of soil samples, an international interdisciplinary team of researchers writing in *Frontiers in Ecology* in 2016 found "an existing, yet overlooked soil management system that has long been—and continues to be—an important feature of the indigenous West African agricultural repertoire" that "transforms highly weathered, nutrient- and carbon-poor tropical soils into enduringly fertile, carbon-rich black soils" through the use of recycled sewage. Chinampas are floating gardens first created by the Aztecs. Reeds and local vegetation are used to raise land inside a shallow lake bed. The mud at the bottom of the lake is rich in nutrients, and the vegetation contributes as well. The soils—compared with their uncultivated peers—are rich in bacteria. Centuries ago, within these systems, corn, tomatoes, squash, chili peppers, and other plants were fed nutrients by bacteria, and they thrived. They supported the creation of Tenochtitlan, modern-day Mexico City. As the Aztec empire expanded, its system of chinampas expanded with it. Since the 1970s, chinampas have been rebuilt. In Tabasco today, some members of the Chontal tribe raise crops through the use of chinampas and enjoy food security as a result.

In the borderland between Laos, Thailand, and Vietnam, in the mountainous headwaters of the Mekong River, where, traditionally, farmers learned to measure the acidity of the soil by tasting it, nutrients in rice paddies are increasingly preserved through varied systems of intercropping trees. Recently, following the introduction of pada trees at a frequency of about one tree per 270 square feet, rice yields were shown to double. At the edge of the region, in modern-day Yunnan, China, the Hani community has long subsisted by adhering strictly to forest conservation, cultivating rice in wetlands, and growing cash crops, including ginger, upland. In 2008, a group of Thai agricultural researchers from Chiang Mai University detailed the conservation practices of the small northern Thai village of Pah Poo Chom, where water is protected as a communal asset and a large forest is managed for its utility "to provide products from timber to wild vegetables and mushrooms that are no longer available in the natural forest." Given these examples and others, the authors wrote, "it is now difficult to imagine that small farmers were once viewed in scientific literature as . . . resistant to change, with their practices looked down upon as archaic and outdated."

In Nauru, refugee camps were built where sustainable agriculture had once been practiced, and the people in those camps were fed with food imported from Australia. However, despite the loss of sustainable practice, local knowledge continues to exist. During World War II, when the island was occupied by Japan's military, its residents were isolated from the outside world. It was an awful time for the Nauruan people, many of whom were deported or killed by Japanese soldiers. Those who survived turned to the past, recycling sewage and other wastes in order to grow food in local gardens. "These guano gardens produced a bumper crop of pumpkins, as well as maize, sweet potatoes, and a little tobacco, and kept a population of more than 5,500 alive and in relatively good health until an Australian relief expedition arrived," the historian Gregory Cushman wrote. They survived that way for two years. Between the start of mining on Nauru and its eventual end, those were the only years when phosphate was returned to the ground.

10

TINY TRACINGS ON A FUTURE WORLD

The phosphate that we mine and use does not depart our earth. It passes through our food, our waste, our sewage plants—or disappears beforehand, running off into a river, building up within a pond, a lake, a bay. It cycles there, it causes growth—perhaps for years—and then it washes out, transitions, flows into the sea. The ocean moves in currents, warm water on top and cold on the bottom. Sun-heated water crosses the Pacific, passes Indonesia, makes its way around South Africa, Agulhas, the Cape of Good Hope. It adjusts course northward up the Atlantic. At Iceland, it cools, sinks, and reverses. It hits Antarctica and splits. One branch travels north, rising near India, joining the upper drift. The other branch stays south, rounds Australia, and moves into the North Pacific. Nauru is where the water surfaces, where it warms, comes up for breath. It has not seen the light of day for multiple millennia. It carries with it nutrients that were preserved, phosphorus that has been gone since Leonardo, since Jesus, since Herodotus, Tutankhamen, Hezekiah. Algae, growing in the light, consume the contents of these newly freed waters. Fish eat the algae. Birds eat the fish. Those birds nest on nearby islands, dropping off their feces as reward. That material, preserved, condenses, turns to guano: harvestable, rich. Climbing upward on this ladder of nutrition, phosphate makes its way onto the land.

But concentrations of nutrients never stay put. They are always being shifted, by geology or by life. They are found and moved, broken up, and left somewhere anew. Trade winds carry minerals across the continents. Dust from Asia makes its way to Hawai'i; it continues on to California every year. The chemical composition of desert sand varies, but some of it contains phosphate. The mountains around Yosemite are kept fertile by Gobi dust, remnants of Gobi sand. Up to half the available phosphorus in the Sierra Nevada came from dust that originated in the lands of the old Mongol Empire. The Bodélé Depression, the lowest point in Chad, is a part of the Sahara that used to be a lake. The ancient lake bed has preserved the remnants of the organisms that once lived in it, and the sand in that region is rich in phosphate. Every year, dust from the Bodélé rises and flows across the Atlantic to the Amazon rainforest, which it fertilizes. The massive outflow of the Amazon River—it provides a fifth of the fresh water that enters oceans worldwide—washes away large amounts of phosphate each year. The soil of the rainforest, perhaps surprisingly, is nutrient poor, because the nutrients are held by plants, not by the ground. The phosphate dust of the Sahara fulfills a need: it replaces the phosphate the river takes away, keeping the ecosystem in balance.

Humans took the phosphate that had built up on the islands, in the soil, beneath the ground. We used it and released it. We wondered about its cause. Buckland saw in buried phosphate "records of warfare, waged by successive generations of inhabitants of our planet on one another." Jenyns described how "in these little stones" we "see the tracings of the fingers of God." Geologists pondered why so much richness had accumulated in one place, why the path of phosphorus had changed. The answer was that certain times were richer for life on earth—that mountains grew more quickly, or the composition of the atmosphere changed, or one of any number of other geophysical processes occurred to expose more phosphate to the living world. On each occasion, phosphate rock became a monument to its time. It marked a period of incredible prosperity, driven by outside forces that fostered life.

So what now? Hearing of these great prosperities of the past, these profusions of global phosphate that created richer times, my mind drifts to the present. Is it possible, as humans ponder the conditions that created phosphate rock, that we are re-creating those conditions in the world today? We have undoubtedly changed the path of phosphorus. We have done the work of mountains or of shifting continental plates—we have enriched the earth. The role of life is to concentrate phosphorus, and humans have carried out that role to a staggering degree. Of the 2.7 pounds of phosphate that the average person consumes annually, more than half—56 percent—originated in a mine. Where will all this phosphate go? Will it re-create the vast deposits that were formed before, leaving wealth below the ground? After millions of years, will strange collections of phosphorus be held beneath the surface? Will it be possible, in the future, to look back at the present day, just as Henslow studied the London Clay, and ask, Why was all this phosphate left behind?

Phosphate rolls off into rivers, which return it to the sea. The waters of the world are accumulating fertilizer lost through the processes of farming and waste. In Iowa, where corn is grown with phosphate mined in Florida, manure from animal factories pollutes rivers. In North Carolina, a center for the hog industry, vast waste pits holding manure are known to flood, particularly during hurricanes, infecting drinking water and fomenting disease. Rivers along the Queensland coast in Australia release four times their normal rate of phosphorus, feeding algae that suffocate portions of the Great Barrier Reef. In India, more than three-quarters of phosphate fertilizer used on farmland comes from mined rather than organic sources, even as eutrophication of coastal waters, fed by unused animal waste, has intensified. State subsidies of phosphate rock–based fertilizers in India make it cheaper to buy fertilizer and leave manure to run off into streams.

In a eutrophic lake or pond or bay, the oxygen is gone, and the creatures that breathe oxygen have disappeared, too. The water is often thick with rotted algae and the bodies of suffocated fish. On the continents of Asia, Europe, and North America, the majority of lakes have become eutrophic. The Baltic Sea is eutrophic. The Chesapeake Bay is eutrophic. Parts of the Gulf of Mexico are eutrophic. In 2001, China, Thailand, and Vietnam collectively raised and killed a majority of the world's hogs. The industrialized facilities where the hogs were raised released large amounts of manure into rivers that emptied into the South China Sea. The South China Sea is eutrophic. In China, as a whole, the amount of manure reused as fertilizer declined from 100 percent to 50 percent between 1950 and 2010, as the amount of organic waste produced by animals increased by more than a factor of five—meaning that 2.5 times as much manure is wasted today as was produced in total (and used) each year in the middle of the past century. Of sixty-seven major inland lakes in China, forty-nine had eutrophied by 2016. In 2008, the marine scientists Robert Diaz and Rutger Rosenberg calculated that the global count of dead zones had doubled every decade since 1960.

Something like this has happened before. There were times in history when phosphate was released from land at elevated levels, when water turned thick and oxygen disappeared. These were the times—these occasional episodes of mess and mystery—when phosphate rock was formed. There were geological causes for these events. A mountain range quickly grew and weathered, continents shifted, the oceans' currents changed. The breakup of Pangaea exposed the edges of continents to erosion, releasing phosphorus into the oceans. The Himalayas' growth spurt did the same. The waters were awash in organic matter, in decomposing bodies, and they experienced periods of oxygen deprivation. Those times were typified by what people today would call dead zones: areas of eutrophication. During the Toarcian age, 183 million years ago, large portions of the earth's oceans had their oxygen sucked out, rendering them anoxic. It happened again in the Aptian 63 million years later, and at the Cenomanian-Turonian boundary 27 million years after that. During the Paleocene-Eocene Thermal Maximum, 56 million years ago, warm temperatures increased the breakdown of mountains yet again. Phosphorus rolled with the waves, and anoxic conditions prevailed.

These events left behind phosphate in the fossil record. This is the phosphate that was mined.

"In the end, phosphorus always ends up in the sediment somewhere. Rivers are the input. Burial in sediments is the sink. There's no other way of getting out," the geologist Caroline Slomp, who studies sediment dynamics in the Baltic Sea, told me. The Baltic, a young sea, is a place that stores its history. Eleven thousand years ago, it was an ice sheet. Then, for three thousand years after melting, it was not a sea at all, but a lake. Eight millennia ago, the lake opened up to the ocean. Throughout the centuries that followed, the Baltic became the home of the Vikings, the beating heart of the Hanseatic League, and the birthplace of the global shipping industry. The entire Baltic Sea has frozen, end to end, twenty times in the last three centuries. The seafloor has been rising ever since the ice sheet disappeared eleven thousand years ago. Rising with it are countless ships and aircraft, downed in global wars and held in freezing, saltless water at the Baltic's bottom. There are shipwrecks spanning centuries; they do not disintegrate. They preserve the lasting power of the sea.

Between 1900 and 1990, as the use of phosphate fertilizer increased, the total annual flow of phosphate into the Baltic Sea grew by a factor of eight. In the final decades of the twentieth century, a dead zone in those waters became the largest in the world. In 1974, the countries that border the Baltic signed the Helsinki Convention, an agreement that was intended to put in place measures to reduce eutrophication in the sea. The convention established the Baltic Marine Environment Protection Commission, a regional policymaking organization. In 2007, the signatories to the Helsinki Convention adopted the Baltic Sea Action Plan, which included forty-two eutrophication-related actions, most of them connected to wastewater or agriculture. In 2021, when the commission reported that eleven of its eutrophication actions had been accomplished, an algal bloom covered more than a third of the surface of the sea. The southern third of the Baltic generally lacks oxygen on the seafloor. The ecosystem has collapsed. Researchers estimate that 30 percent of Baltic secondary production—fish and other organisms that consume plants—has disappeared as a result. "If the dead zones were eliminated, the Baltic would be more productive by at least a third to a half," Diaz and Rosenberg, the marine scientists, wrote.

On land, in natural spaces, phosphorus cycles through living bodies an average of forty-six times before it travels to the sea. In the sea, it passes through, on average, eight hundred different bodies before it drifts out of the biological world and into the geological one. The residence time of phosphate in the Baltic Sea is likely several decades. The phosphate flowing now will remain there for the foreseeable future. It will feed new organisms, decompose, and foster bacteria. It will continue this cycle every year. Far in the future, following hundreds of biological cycles crossing the course of several decades, the phosphate in the Baltic Sea will settle, and the layer of sediment that forms will be richer in phosphorus than it otherwise would have been. Vivianite, an iron-phosphate mineral, is one example of what might be left behind by these events. Vivianite forms from water in places where phosphate levels are high. It is bright blue. It has been found around the world in bogs and lakes, in excrement and decomposing leaves, on rotted roots and dead human bodies, within which it replaces teeth and bones. In 2015, Slomp, the geologist, led a team of researchers that discovered evidence of growing vivianite crystals in the Baltic.

For eleven years, the geologist Jan Zalasiewicz chaired the Anthropocene Working Group, the international organization of geologists working to collect evidence of the formation of a new, human-dominated epoch of earth history. This proposed new epoch is defined by the increased emission of carbon into the atmosphere, by the proliferation of plastics, by the growth of major cities and the rising levels of the sea. Another legacy of humans may be the mass movement of phosphate from land to water. Today, as anoxia increases again, the earth is coming closer to the way it was during the Cretaceous: temperatures are warming, oxygen levels are diminishing in the water, and oceanic circulation may be slowing down. There is a question of whether large-scale deposition will repeat itself. "At the beginning of the Cambrian period there is a kind of phosphate event. And there are others going through geological history, of times where strata seemed to have more phosphate than usual in them. One query I have in the back of my mind is whether we are creating another such, if you like, phosphate stratigraphic layer or event through our digging out so much of it and spreading it out on the surface, whether that is significant. We know its environmental effects—the creation of dead zones and so on—but if you're looking through to the geological future, to see what is going to be the end result after the phosphate perturbation, it's something which I think has not been studied very much," Zalasiewicz told me. The accumulation of phosphate, the collection of clues that might provide an answer to that question, occurs on the seafloors of dead zones, where researchers are looking for indications of freshly growing phosphate rock—"the repository," Zalasiewicz said. Rich with phosphate runoff, coastal waters are losing their oxygen. Our planet may be reverting to conditions that led to phosphate deposition in Morocco, in England, in Florida long ago. We are burying phosphate anew.

In the very long term, phosphorus deposition generally corresponds to phosphorus flux. That is, over a period of, say, one hundred thousand years, an increase in phosphate in the sedimentary record is a good indicator that something was happening on land. In 1992, the geochemist Kathleen Ruttenberg developed a method, called SEDEX, for quantifying phosphorus present in sediment. She was studying the floor of the Mississippi basin. Her paper describing the SEDEX method has been cited more than 1,300 times by researchers who have observed the modern formation of phosphate rock all around the world—off China, off India, around Europe, and in the Mediterranean. There is evidence of phosphate rock formation in the sediments along the Mexican continental margin. Phosphate rock is forming in the Santa Barbara Basin and Long Island Sound. Vivianite, found in the Baltic, is also growing in the Chesapeake Bay, offshore of the Amazon, on the Yung-An Ridge in the South China Sea, and in the Bothnian Sea. Specimens of vivianite have been found in lakes in Norway, Germany, Italy, the United States, Canada, Japan, Sweden, Russia, Argentina, France, Denmark, and the United Kingdom. This phosphate will likely be visible in the geologic record; the strata will bear the remnants of superphosphate, of Lawes and Liebig and Borlaug and millions of farmers planting over centuries. A future Henslow walking along a newly rediscovered cliff may find nodules, one day far from now, and wonder about what life-forms created them. These features, present in the guarded water bodies of the earth, will be markers of our time.

Humanity is creating a record of change, but that record will not be stable around the world. In some places, including the Baltic Sea and the Chesapeake Bay, phosphate is collected and held; it will be visible in the future. But the large bulk of phosphate that is not stored in mountains or lakes or plants or people will find its way into the deepest parts of the ocean. The chemical situation of phosphate in the ocean varies widely. The vast majority of it is inaccessible, nonreactive, bound up in crystal structures that generally do not break. Much of the reactive phosphate is bound to iron oxides—rust—that can only be torn apart under the right conditions. Large amounts of phosphate are contained in organic matter: dead fish, decaying plants, old algae. Pieces of organic matter that form in upwelled waters do not make it onto land. They sink and collect along the continental shelf. They condense, and, sometimes, they turn to apatite, phosphate rock—the same material mined in Florida and Morocco today. "You need exceptional areas of upwelling to get the gooey, oozy, organic-rich sediment required for phosphate rock to actually start to form," the geologist John Compton said.

In the oceans, the effect of humans will often be unclear. Our heightened phosphorus use will be time-limited; we are unlikely to produce enough to leave a cohesive layer around the world. "I'm not convinced that we'll necessarily always see it," Slomp, the Baltic geologist, told me. "Ultimately, in the long run, what we're putting in from fertilizer isn't enough—we can't sustain that for a long time." The geologist Gabriel Filippelli placed the exhaustion of phosphate resources at 3600 CE, a year that is by no means set in stone. Other geologists have placed the date at various points earlier, but that exhaustion is beside the point: farmland is saturated with phosphate fertilizer, which is working with declining effect. Recognition of environmental harm is increasing. The use of phosphate rock is no longer increasing. Since the nineteenth century, the flow of phosphate from land into the oceans of the world has doubled due to human activity—but that doubling amounts to only a small fraction of the phosphate already present in the deep sea. The timescale on which the ocean operates is far greater than the one in which we live. The ocean is a vast place. Worldwide, our effect may amount to just a bump in the graph.

The two deposits most analogous to the ones we mine today—the two places in the world where Florida and Morocco are being replaced—are growing off the coastlines of Peru and Namibia. Ancient phosphate is coming from the ocean floor, and it is settling on shelves along the continents. In places that lack oxygen, the particles break apart, leaving phosphorus to crystallize into rock. Sometimes, the right conditions are created by bacteria, which tend to cluster, forming tiny blobs made up of phosphorus. "These bacteria are taking the phosphorus that's coming from up above—maybe some coming from iron oxides, some from the decay of organic matter—they're sitting in the perfect zone, and they accumulate a ton of phosphorus within their cells," the geologist Ellery Ingall, who studies these organisms, told me. "Conditions change. If oxygen goes away, if there's an earthquake that buries that mat, now you have this layer that's really, really rich in phosphorus. And that can trigger precipitation." Typically, phosphate rock forms somewhere in the sediments. Tiny crystals are buried by mud. Atop the mud, organic matter is thickly layered. Over millions of years, that organic matter is taken by the concentrated phosphorus: it is trapped within the crystals of the newly forming rock. Phosphate rock extends throughout the layer. "Years ago, I did an expedition down to the Peru shelf in a submersible. You could see these bacterial mats that tend to concentrate phosphorus—just this layer of gelatinous material going for miles on the ocean floor. In other places, there were phosphate hardgrounds. There was so much phosphorus that bones or anything that fell on the seafloor would be phosphatized. There were layers of fish bones that were, in reality, hard phosphate rock. I was impressed by the sheer extent of some of these bacterial mats. When we would sample them with a little core tube, working it in with the arm on the submersible, they were like Jell-O, and the whole mat would shake for several centimeters around. Those mats are probably ninety-nine percent water. We could squeeze those mats and look at the dissolved phosphorus coming off of them," Ingall said.

The earth will envelop this into its system. In some places, in the future, the effects of modern agriculture will be visible. There will be vivianite in the Baltic Sea, vivianite in the Chesapeake Bay, other kinds of phosphate rock forming in the places where humans exerted considerable effect. There will be phosphate rock formations elsewhere, too, and they may be marginally larger because of us. Phosphate rock on land will be reduced, but new deposits will be formed in the sea. Some of those will be because of us; others will result from the unchanged cycles of the earth. (Phosphate would be forming off Peru even without the influence of humanity, though our effect may ultimately increase the size of that deposit.) Some phosphate, floating in the seas, will turn to sediments. Those particles will become a part of the construction of the slowly growing ocean floor. Hundreds of millions of years into the future, the crust of the ocean will subduct beneath the continents. It will dissipate within the mantle, where reservoirs of elements are kept. Eventually, that phosphate will become part of the lava that erupts from volcanoes, part of the rock that forms at rifts between the plates. It will re-create the structures of the surface of the earth. It will erode. It will live again.

Eight and a half thousand miles separate Nauru, where something ended, from England, where it began. It is an arbitrary start and finish: we were always taking phosphate from the ground, and we always will be. But a special hope arose at Bawdsey in 1842, a special cynicism in Nauru a century and a half later. That hope, and that cynicism, tell us something of how we all have changed.

What do we make of the remains—of the degradation of farmland, of the artificial richness of the sea? Of disappeared bodies that have turned to floating gunk? Of solitary crystals? The layers, the striations, the conflicting angles of the ground have told us stories of the changes of the past: currents, tidal movements, and storms.

Particles tumbled and settled, becoming phosphate rock. They were liberated by violence, recast and rediscovered. The alchemist, it has been written, "contributes to Nature's work by precipitating the rhythm of Time." Most of us succumb to Time. We cast judgments. We have piles of phosphogypsum strewn in perpetuity. The quality of our vegetables has plummeted. We torture animals—dreadfully, wantonly, without a long-term benefit. (How can anybody sleep at night?) "Life changes," someone said, speaking of phosphorus, "as you get richer and richer."

It was possible, when Lieutenant-Colonel James Langley visited Nauru in the 1960s, to place a phone call back to his associates in England. Langley had been a hero in the Battle of Dunkirk, during the Second World War. Wounded, captured, and subsequently freed, he became a spy for England. After the war, he went to work for Fisons, traveling the world. The ship he took to the Pacific could hold twelve thousand tons of phosphate. He wrote, "It was from Nauru that I telephoned to Suffolk—undoubtedly one of the longest telephone calls that can be made."

The phosphate rock of Nauru had already gone to Suffolk, home of Henslow, of the Bawdsey Cliff, of Packard's and a history of fire. The reality of eutrophication was essentially clear. The failures of the green revolution had been foretold. The phosphorus cycle was long since ruptured—Liebig, creating fertilizer, had demonstrated that. Liebig's writings shaped the views of Marx and Mill, of robbery, of metabolic rift. Marx only outlived Liebig by a decade. Did the alchemist contribute to nature or detract from it, when he sped up Time?

Oceanic currents encircle the world, and through them, life has changed. We fly airplanes filled with phosphorus we drop as bombs, sometimes as fertilizer. We marvel at it. The mystery of phosphorus was not so different from the question of God: Phosphate rock was formed, but why? It grew in dunes, in waves, but why? Why was it *here*? As soon as Henslow found the rock, geologists grasped the conditions that had created it. The mechanism presented itself, but the cause did not. Perhaps, this time, we ought to leave a note. *We did this,* it will say. *These are our lives you found.*

In Suffolk, the world saw promise in phosphate rock. In Nauru, it sees a parable of defeat. Even Langley understood what brought these places together. (None of this is new.) A five-minute call, routed through Sydney, made up of words he did not leave for us to read. They were probably banal. Fisons was on the line. This was a call to the beginning from the end.

In the moments that follow the death of a whale, a world is born, and then the earth erases it. The hagfish come, the sharks, bacteria. We all feed on what we have been given, blindly searching for something in the dark. We find it. "It is," an observer said, "like the enchanted Isle of Monte Christo or the wealth of Croesus." We have little inkling of our context.

Our peaks are life-defining and inconsequential. We are everything and nothing. We have altered our connection with the earth. The manner of our bond is broken—we have isolated and worshipped it, taken its essence as our property. And yet the fundamentals of our planet overwhelm us. We cannot compete with time. Discoveries will continue apace: the tree's discovery of phosphate leaking from a stream; the beetle's realization of dung on which to perch; the hagfish, nasty but wise, that finds a whale beneath the waves; the rector, pedigreed, collecting stones along a cliff. It is our connection, our eternity.

Hennig Brand, toiling away at destiny, really did discover the philosopher's stone: through the light of phosphorus, we live forever. We regenerate, and our parts create new wholes. The world warms, and life surges; it builds upon itself. Dust blows, currents shift, continents move, species come and go. *The blood is the life,* Moses said, and people listened, giving it to the ground in offering, and they reached the Promised Land. But they were thirsty.

"Take the rod," God said to Moses, "and speak ye unto the rock before their eyes; and it shall give forth water." But the rod was a temptation, and God was testing Moses's faith. Moses needed to trust in God—to speak, as God had said, and not to hit the rock.

"And Moses lifted up his hand, and with his rod he smote the rock," the Bible tells us, "and the water came out abundantly."

God was angry. "Because ye believed me not," he said, "therefore ye shall not bring this congregation into the land which I have given them." The Israelites were banished from the Promised Land, cast back out into the desert, and they searched for sustenance again. Their prosperity had lasted only decades.

As a garden follows the seasons, so a galaxy dies in time. The stars were born, they grew like herbs, they populated space. The skies were spinning and expanding. "At all events, it was clear that the Milky Way 'cannot last forever'; and equally that 'its past duration cannot be admitted to be infinite.' It followed that neither the earth, nor even the solar system, was a separate creation, but merely an infinitesimal part of a galactic evolution. Our galaxy had a physical beginning, and would have a physical conclusion. Our solar system, our planet, and hence our whole civilisation would have an ultimate and unavoidable end," wrote Richard Holmes, the eminent modern historian of the Romantic era, of William Herschel, the astronomer who died in 1822. Herschel's was a progressive view: others saw astronomy as a tool for navigation and mathematics. But for Herschel, who emigrated from Germany and built telescopes in the English countryside, astronomy was the study of the evolutions of the skies, of the pathways of time. All astronomy was narrative. It was romantic, though he was cold.

In 1781, Herschel discovered the planet Uranus. Two centuries later, a telescope was built in the Canary Islands, atop the Roque de los Muchachos, a craggy pinnacle that reaches toward the sky. It was 1987. The telescope was named for Herschel. Thirty years after its opening, the William Hershel Telescope played host to a certain search for life. Astronomers were looking for phosphorus in the heavens. They took pictures of the materials left by declining supernovas, seeking evidence of material that creates life. They believed that signs of phosphorus would be the surest indication that a faraway solar system could hold biology. In 2018, they announced preliminary results of their study. They had found no evidence of phosphate far away; within our galaxy, it seems, earth is lucky and alone.

So here we are. We are flecks of dust on the curtain of the cosmos. We are solitary animals, crouched atop a rocky world, feeding from stones that roll off mountaintops. We are made of death, we die, and then we come alive again. We search for pieces of prosperity and we grow profusely, feasting on the souls of former beings. We dazzle and revel and burn with fires of creation, but our everlasting afterlives require constant death. The universe, like the deep sea, is dark, distant, and expansive—and on the seafloor, every whale fall one day ends.

ACKNOWLEDGMENTS

Where to begin? I am grateful to George Lucas, who supported this book from the start, and Todd Portnowitz, who guided it expertly for the past few years. The publishing process included many great people, including Ben Shields, Amy Ryan, Kathleen Cook, Lisa Lucas, and Linda Huang. I am grateful to all of them.

Alice Maiden, the illustrator of this book, was also its first reader. The images she created not only captured the book's spirit—they shaped it. I am thankful to Jonah and Anika for their perceptive literary feedback and also, more importantly, for their friendship through the years. A list of those wonderful people who cared for me and kept me going would be endless, although it would include Odonoo, for many reasons, and, of course, my parents and my sisters. Thank you.

To speak to a writer who intends to print your name and facts about you requires, I am sure, a great amount of open-mindedness. If I interviewed you for this book, please know that I am grateful for your time, your trust, and your wisdom. I would like to thank some people in particular: in England, Liz Harper, Neil Davies, Bob Markham, Simon Jackson, Gerald Tarbet, and Kelvin Dakin; in Florida, Thom Downs, Norma Killebrew, Paul DeGaeta, Andy Mele, Barbara Fite, and Jeff Barath; in Melilla, Muhammed Ahl Abhayah, among others; in Australia, Roland Kun, Julian Burnside,

Geoffrey Eames, and Peter Law; in New Zealand, Tracey Barnett, Becs Amundsen, and the late Sue Morrison-Bailey; in Kiribati, Pelenise Alofa; and in Nauru, my dear friends Kathy Barazandeh and Nanise Keni.

I am not a specialist in any of the disciplines covered in this book, and so I am deeply grateful to the many people who proofread chapters or talked me through complicated issues of economics, history, ecology, biology, chemistry, soil science, and, of course, geology. This group of people includes John Ikerd, David Vaccari, Harvey Osborne, Dan Canfield, Stuart White, Martin Blackwell, Susan Tandy, Andy Neal, Melanie Wright, Nigel Woodcock, Eric Hiatt, Ellery Ingall, Henrik Schiellerup, Nick Butterfield, John Compton, Jan Zalasiewicz, Caroline Slomp, and many of the people named above. Bernard O'Connor, working independently, conducted extraordinary research into the early history of phosphate mining in England, and he generously shared that work with me. Corey Dickinson, many years ago, provided a multitude of personal materials relating to Nauru.

In researching this book, I was aided by talented experts working in a variety of fields at a number of libraries and archives around the world. These institutions included, at the University of Cambridge, the Cambridge University Library, the Munby Rare Books Reading Room, the Earth Sciences Library, the Sedgwick Museum, and the libraries and archives of the following colleges: Christ's, King's, St. John's, Trinity, St. Catharine's, and Queens'. (Genny Silvanus, of Christ's College, was a particularly wonderful collaborator.) In Ely, the Cambridgeshire Archives. In London, the National Archives at Kew and the Valence House Museum. In Florida, the Florida Industrial and Phosphate Research Institute. The public library in Nauru was invaluable, as was the Firestone Library at Princeton University.

This reporting was funded by grants from a collection of institutions within Princeton University: the Princeton Environmental Institute, now called the High Meadows Environmental Institute (Smith-Newton Environmental Scholars Program; Environmental Senior Thesis Prize), the Department of English (Maren Grant; Charles R. Kennedy Prize), the Program in Journalism (John

McPhee Award for Interdisciplinary Reporting), and the Princeton Institute for International Research (2018 fellowship). Multiple professors at Princeton supported this book. Tamsen Wolff and Rob Nixon advised my work on Nauru, which is how this project began. Diana Fuss, Pico Iyer, Joe Stephens, and John McPhee all provided help along the way. They believed in me and encouraged me to believe in myself.

Before I wanted to be a writer, I wanted to be a farmer, and it was my grandmother, Mary Little, who taught me to compost, to grow flowers and vegetables, and to love the earth.

NOTES

PROLOGUE: WHALE FALL

x "life's bottleneck": Isaac Asimov, "Life's Bottleneck," in *Asimov on Chemistry* (New York: Anchor Books, 1974), 160–70.

xiii the mix of chemical elements: Stephen Porder and Sohini Ramachandran, "The Phosphorus Concentration of Common Rocks—A Potential Driver of Ecosystem P Status," *Plant and Soil* 367, nos. 1–2 (2013): 41–55. On a very rough average, granite contains 440 parts per million of phosphorus, an amount equal to just under 0.05 percent of the rock's mass. Limestone and marble contain 570. Quartzite—which is metamorphic—contains about 340 parts per million of phosphorus, while slate contains 600. Sandstone, a common sedimentary rock, holds 500. The composition of rocks is highly variable, and these numbers, the averages of data sets with very high ranges, derive from a study based on measurements of 263,000 rock samples.

CHAPTER 1: SEA OF FIRES

7 particularly dire: John E. Archer, *Social Unrest and Popular Protest in England, 1780–1840* (Cambridge: Cambridge University Press, 2000), 10–11; John E. Archer, *By a Flash and a Scare: Arson, Animal Maiming, and Poaching in East Anglia, 1815–1870* (Oxford: Clarendon Press, 1990), 137–41.

7 The textile economy had declined: This was explained to me by Harvey Osborne, of the University of Suffolk, who also shared important background information about Hitcham, Henslow, and the crisis of the 1840s.

7 more than six hundred fires: Archer, *By a Flash and a Scare,* 176.

7 "the match of the prowling incendiary": S. M. Walters and E. A. Stow, *Darwin's Mentor: John Stevens Henslow, 1796–1861* (Cambridge: Cambridge University Press, 2009), 197.

7 "Two preachers": Archer, *By a Flash and a Scare,* 108–9.

8 the parsonage had been set aflame: Archer, *By a Flash and a Scare,* 143.

8 "a cry of fire interrupted": Walters and Stow, *Darwin's Mentor,* 197.

8 *The Times* of London reported: Archer, *By a Flash and a Scare,* 109.

9 "No farmer felt himself secure": Leonard Jenyns, *Memoir of the Rev. John Stevens Henslow, M.A., F.L.S., F.G.S., F.C.P.S.: Late Rector of Hitcham and Professor of Botany in the University of Cambridge* (London: John van Voorst, 1862), 88.

9 schemes for the poor: Jean P. Russell-Gebbett, *Henslow of Hitcham: Botanist, Educationalist, and Clergyman* (Lavenham, Suffolk: Terence Dalton Limited, 1977), 33–34.

9 founded a school: Russell-Gebbett, *Henslow of Hitcham,* 31.

9 a series of experiments: Walters and Stow, *Darwin's Mentor,* 200, 211.

9 He lectured on farming methods: Jenyns, *Memoir of the Rev. John Stevens Henslow,* 77–82.

9 He gave lessons on manuring: Russell-Gebbett, *Henslow of Hitcham,* 42, 83.

10 "The *art* of husbandry": Jenyns, *Memoir of the Rev. John Stevens Henslow,* 82.

10 Henslow took a walk: Jenyns, *Memoir of the Rev. John Stevens Henslow,* 201–3.

11 became dry land: J. R. Lee, M. A. Woods, and B. S. P. Moorlock, *British Regional Geology: East Anglia,* 5th ed. (Nottingham: British Geological Survey, 2015), 89.

15 coprolites: From the beginning, geologists probing the history of English phosphate discovered these connections between the status of the world and the preservation of its phosphate. In 1853, the Scottish agricultural chemist James Finlay Weir Johnston proposed an origin story for the phosphate in the Crag: "that the phosphorus had been derived from animal remains in higher strata, dissolved out by acids, and redeposited at a lower level." Searles Valentine Wood, a paleontologist who had worked as a midshipman, speculated that the phosphate had been reworked by tidal currents, a theory that is known today to be true. Wood imagined a sea that lapped at the London Clay, shifting its fossils and burying them under successive layers of sediment. Throughout the 1860s, Edwin Ray Lankester, destined to fame as a zoologist, analyzed the phosphate of the Crag and demonstrated that it derived from animals that lived not all at once but variously over a period of millions of years that spanned the Pliocene, Miocene, and Eocene. Joseph Prestwich, meanwhile, excavated a Crag pit and described an antediluvian world of rising and falling seas—"a shallowing of the sea by a reverse movement of elevation, and the setting in of stronger currents with intervals of quiet deposition"—that created phosphate rock. In 1990, the geologist Peter Balson, who surveyed the east of England during a long career with the British Geological Survey, connected the preservation of phosphate in the Red Crag to global changes in the world's oceans, including the fluctuation of temperatures and the rise and fall of the sea. "Phosphorite concretions are not randomly distributed," Balson wrote in the paper, part of a special BGS publication. "They occur only during specific relatively narrow intervals within marine sequences and are associated with other features diagnostic of open connections with the oceanic realm." Nick Butterfield, a geobiologist at Cambridge, poured out bags of phosphate nodules on a table in his office above the Sedgwick Museum when I asked about the origin of the phosphate that developed nearby. "Is it about input or output?" he said. "Maybe there were massive amounts of phosphate being added to the environment. Maybe plants and fungi were weathering the bedrock and releasing phosphate into the water, and all that phosphate had to go somewhere, and there was a phosphogenic event. Or maybe it's about outputs—animals collected the phosphate in their bodies, and the conditions just happened to be right for them to phosphatize. That's the mystery."

16 "In October, 1843, Prof. Henslow": John Stevens Henslow, "On Nodules, appar-

ently Coprolitic, from the Red Crag, London Clay, and Greensand," *Report of the British Association for the Advancement of Science* (1845): 51–52.

16 "It was in the summer of 1842": John Stevens Henslow, July 3, 1842, letter, *The Gardeners' Chronicle and Agricultural Gazette* (March 11, 1848): 180–81.

16 "The opening of a new pit": Walter Tye, "Coproliting in the Deben Peninsula: Tales of a Once-Thriving Industry That Is Now Almost Forgotten," *Suffolk Chronicle and Mercury* (December 30, 1949): 3.

17 range of ten thousand to twelve thousand tons: A. N. Gray, *Phosphates and Superphosphate* (London: E. T. Heron & Company, Limited, 1944), 21.

17 325,000 tons: Trevor D. Ford and Bernard O'Connor, "A Vanished Industry: Coprolite Mining," *Mercian Geologist* 17, no. 2 (2009): 99. Also see: "The Origins and Development of the British Coprolite Industry," *Mining History* 14, no. 5 (2001).

18 Henslow became president of the Ipswich Museum: Jenyns, *Memoir of the Rev. John Stevens Henslow,* 154.

18 "'like granite wearing away'": Jenyns, *Memoir of the Rev. John Stevens Henslow,* 251.

19 "virtuous rusticity": Archer, *By a Flash and a Scare,* 65.

19 "the fingers of God": Quoted in Bernard O'Connor, *The Suffolk Fossil Diggings* (self-published, 2009), 190–91.

CHAPTER 2: THE ACID TEST

22 one-quarter of the nutrients: Gabriel M. Filippelli, "The Global Phosphorus Cycle," *Elements* 4, no. 2 (2008): 92.

24 "the quagga mussel": Jiying Li et al., "Benthic invaders control the phosphorus cycle in the world's largest freshwater ecosystem," *Proceedings of the National Academy of Sciences* 118, no. 6 (2021): 1–9.

25 modern quinoa: Lawrence A. Kuznar, "Mutualism Between Chenopodium, Herd Animals, and Herders in the South Central Andes," *Mountain Research and Development* 13, no. 3 (August 1993): 257–65. This research was explained to me by Randy Haas, at U.C. Davis.

25 Amazonian Dark Earth: Charles C. Mann, *1491: New Revelations of the Americas Before Columbus* (New York: Knopf, 2005), chapter 9.

26 Pre-Inca civilizations in South America: Pedro Rodrigues and Joana Micael, "The importance of guano birds to the Inca Empire and the first conservation measures implemented by humans," *International Journal of Avian Science* 163 (2021): 283–91.

26 borders of that empire: Rodrigues and Micael, "The importance of guano birds to the Inca Empire," 284.

26 Early civilizations along the Indus River: Eduardo Garzanti et al., "Petrology of Indus River sands: A key to interpret erosion history of the Western Himalayan Syntaxis," *Earth and Planetary Science Letters* 229, nos. 3–4 (January 15, 2005): 287–302.

27 the Ethiopian Plateau: Eduardo Garzanti et al., "The modern Nile sediment system: Processes and products," *Quaternary Science Reviews* 130 (2015): 19. The intricacies of the Nile River basin were explained to me by Michael Krom, of the Israel Oceanographic and Limnological Research Institute.

27 "so little labor": Herodotus, *The Histories,* book 2, trans. A. D. Godley (Cambridge: Harvard University Press, 1920), chapter 14.

27 Theophrastus ranked: Theophrastus, *Enquiry into Plants,* book 2, trans. Arthur Holt (Cambridge: Loeb Classical Library, Harvard University Press, 1916), 148–49.

27 Columella, the Roman agriculturist: Columella, *De re rustica,* book 2, trans. Harrison Boyd Ash (Cambridge: Loeb Classical Library, Harvard University Press, 1941), paragraph 14.

27 "large dunghill": Cato the Elder, *De agri cultura,* trans. W. D. Hooper and Harrison Boyd Ash (Cambridge: Loeb Classical Library, Harvard University Press, 1934), section 5.

27 *laetamen*: Nicola Gardini, *Long Live Latin: The Pleasures of a Useless Language,* trans. Todd Portnowitz (New York: Farrar, Straus and Giroux, 2019), 69.

27 "to manure the king's estates": Homer, *The Odyssey,* trans. Robert Fagles (New York: Penguin Classics, 1996), book 17, lines 225–27.

28 Aryans living near the Ganges: George Ordish et al., "Origins of Agriculture: The Indian Subcontinent," *Encyclopaedia Britannica* (accessed online).

28 The Egyptians viewed dung beetles: "Country Showcase: Ancient Egyptian Agriculture," Food and Agriculture Organization of the United Nations (published online December 6, 2020).

28 In Mesopotamia, the act of composting: M. Patchaye et al., "Microbial management of organic waste in agroecosystem," in *Microorganisms for Green Revolution,* vol. 2, *Microbes for Sustainable Agro-Ecosystem,* ed. D. G. Panpatte et al. (Singapore: Springer, 2018), 51.

28 movement of manure from city to farmland: T. J. Wilkinson, "The Definition of Ancient Manured Zones by Means of Extensive Sherd-Sampling Techniques," *Journal of Field Archaeology* 9, no. 3 (Autumn 1982): 323–33.

28 the ancient Minoan civilization: Andreas Angelakis et al., "Water Reuse: From Ancient to Modern Times and the Future," *Frontiers in Environmental Science* 6 (May 11, 2018): 2–3.

28 ancient Athenians did the same thing: Ellen Churchill Semple, "Ancient Mediterranean Agriculture: Part II. Manuring and Seed Selection," *Agricultural History* 2, no. 3 (July 1928): 139–40.

28 countless biblical examples: Semple, "Ancient Mediterranean Agriculture," 130–31.

28 "blood is the life": Deuteronomy 12:23–25, New International Version.

29 the impurities of feces: Shlomit Paz et al., "The potential conflict between traditional perceptions and environmental behavior: Compost use by Muslim farms," *Environment Development and Sustainability* 15 (2013): 973–75.

29 proponents of the Arab Agricultural Revolution: Andrew M. Watson, "The Arab Agricultural Revolution and Its Diffusion, 700–1100," *The Journal of Economic History* 34, no. 1 (March 1974): 8–35.

29 worm compost ventures in Sudan: Elkhtab Mohamed Abdalla et al., "The chemical characteristics of composted and vermicomposted cotton residues case study in Sudan," *World Journal of Science, Technology and Sustainable Development* 9, no. 4 (2012): 325–36.

29 the smell of a gong farmer's dung cart: John Harington, *A New Discourse of a Stale Subject, Called the Metamorphosis of Ajax* (published by the Ex-Classics Project, www.exclassics.com, 2015; originally published 1596), 29.

29 Soil samples taken around medieval archaeological sites: Elena Chernysheva et al., "Effects of long-term medieval agriculture on soil properties: A case study from the

Kislovodsk basin, Northern Caucasus, Russia," *Journal of Mountain Science* 15, no. 6 (2018): 1176–77.

29 minimal evidence of animal dung accumulation: Dolly Jørgensen, "Managing Streets and Gutters in Late Medieval England and Scandinavia," *Technology and Culture* 49, no. 3 (July 2008): 564.

29 In China, the application: Zitong Gong et al., "Classical Farming Systems of China," *Journal of Crop Production* 3, no. 1 (2001): 11–21.

30 The provision of so-called night soil: Susan B. Hanley, "Urban Sanitation in Preindustrial Japan," *Journal of Interdisciplinary History* 18, no. 1 (Summer 1987): 1–26.

31 "Early representatives of the fertilizer industry": Jan Zalasiewicz, "The Turnip That Ate the Pterodactyl," *The Palaeontology Newsletter* 78 (December 2011): 1–9. Also see: Gordon Mingay, "The sources of productivity in the agricultural revolution in England," *South African Journal of Economic History* 6, no. 1 (1991): 6.

31 thirty thousand tons of bones: Gray, *Phosphates and Superphosphate,* 104–5.

32 "the best bedrooms in the house": John Bennet Lawes and Henry Gilbert, *The Rothamsted Experiments: Results of the Agricultural Investigations Conducted at Rothamsted* (Edinburgh and London: Blackwood and Sons, 1895), 3.

33 In 1835, a German headmaster: Gray, *Phosphates and Superphosphate,* 104–5.

33 English entrepreneurs followed suit: Frank Snowden, "The Sanitary Movement and the 'Filth Theory of Disease,'" lecture 11, Epidemics in Western Society Since 1600 (HIST 234), Open Yale Courses, https://oyc.yale.edu/history/hist-234/lecture-11.

34 Fison was the latest scion: Michael S. Moss, *Fertilizers to Pharmaceuticals—Fisons: The Biography of a Company: 1720–1986* (unpublished), 38, 46. Shared with me by Kelvin Dakin.

34 "separated only by a single wall": Walter G. T. Packard, "Bramford in the 1900s," *Fisons Journal* (Spring 1957): 21.

34 As Packard's, Fisons: Moss, *Fertilizers to Pharmaceuticals,* 59, 61–62, 66.

35 "as a tribute to the town": Moss, *Fertilizers to Pharmaceuticals,* 62.

35 "the eye could not fail to see": George Henslow, "Reminiscences of a Scientific Suffolk Clergyman," *The Eastern Counties Magazine* 1 (1900): 29.

35 "For the privilege of digging": Gray, *Phosphates and Superphosphate,* 21.

35 reached sixty feet in depth: Moss, *Fertilizers to Pharmaceuticals,* 61.

36 72,000 tons of superphosphate: Henslow, "Reminiscences of a Scientific Suffolk Clergyman," 23.

36 During the 1870s: Moss, *Fertilizers to Pharmaceuticals,* 67–68.

36 "The coprolites are trimmed by men": Moss, *Fertilizers to Pharmaceuticals,* 68–69.

36 The Florida Phosphate Company: Moss, *Fertilizers to Pharmaceuticals,* 89.

37 "Algeria, the West Indies": Moss, *Fertilizers to Pharmaceuticals,* 90.

37 Russia, Martinique, and Japan: Moss, *Fertilizers to Pharmaceuticals,* 93.

37 The ledgers of the Lawes Chemical Manure Company: "Minute Book 1" (NR140/1), 194, 239–41, 299, 325, 335; "Minute Book 2" (NR140/2), 14, 26, 27, 52, 53, 95–96, 114, 137, 143, 154, 156, 161, 168–69, 173–74, 180, 187, 191–92; "Minute Book 3" (NR140/3), 12, 16, 17, 23, 42, 80–90, 97–98, 159, 202, 215; "Minute Book 4" (NR140/4), 1, 15, 140, 168, 183, 196; "Minute Book 5" (NR140/5), 262, 286, 297, 302; "Minute Book 6" (NR140/6), 72, 89, 134–35, 158, 195, 226, 282, Archive of the Lawes Chemical Manure Company, Valence House Museum, Dagenham.

38 "a rich and cheap source of phosphate": Quoted in Henslow, "Reminiscences of a Scientific Suffolk Clergyman," 22–23.

38 "various and abundant": Henslow, "Reminiscences of a Scientific Suffolk Clergyman," 23.

38 "In these depositories of the dead": "Catacombs—Great Deposits of Phosphate of Lime," *Scientific American* (December 5, 1857).

38 "Great Britain deprives all countries": Translated from *Einleitung,* vol. 1, 125. Quoted in William H. Brock, *Justus von Liebig: The Chemical Gatekeeper* (New York: Cambridge University Press, 1997), 178.

39 "These are not vague and dark prophecies": Quoted in *The Daily Telegraph,* October 1862 (press cutting saved in Rothamsted archives). Also quoted in William H. Brock, *Justus von Liebig,* 178.

40 "to help us up the sides of Eldey": W. G. T. Packard, "A Wild Goose Chase," *The F.P.P. Journal* 3 (January 1931): 18–25. All copies of *Fisons Journal*—first called *The F.P.P. Journal*—were shared with me by Kelvin Dakin, of the Bramford Local History Group. Many of these documents were originally collected by the late Brian Blomfield.

40 In 1961, when Fisons laid out its growth plans: "Fisons Overseas Limited," *Fisons Journal* 70 (July 1961): 34–37.

41 Bob Markham: Some of the specimens collected by Bob Markham are kept inside the bonery at the Ipswich Museum, which is located beneath a rabbits' warren of interlocking stairs and passageways. Markham was a curator at the museum for thirty years, during which time he safeguarded the collections that Henslow had developed more than a hundred years before. In Red Crag geology, Markham is a legend. "He knows huge amounts about the crag, and a huge amount about the distributions of the organisms that lived there. If you say things to him—you spent a bit of time working something out—you realize that he knows already," Liz Harper told me as we walked around the Fens north of Cambridge. During his career at the museum, Markham accumulated a vast amount of evidence of phosphate formation throughout the east of England. "We estimate we've got twenty thousand in our Crag collection. A lot of it is thanks to Bob Markham's collecting," Simon Jackson, the museum's new geology curator, said.

41 St. John's College: *Rep. Com. Univ. Inc.,* vol. 1, 177, 379, shared with me by Bernard O'Connor.

41 King's College: *Rep. Com. Univ. Inc.,* vol. 3, 198–200, 211–12, shared with me by Bernard O'Connor.

41 Christ's College: "Lady Day Audit 1882," "Michaelmas Audit 1882," "Lady Day Audit 1883," "Michaelmas Audit 1883," Christ's College Archives; *Rep. Com. Univ. Inc.,* vol. 3, 314, 317, shared with me by Bernard O'Connor.

41 Construction had boomed: Data shared with me by Nigel Woodcock.

42 a slice of rock: Rob Clarke, a rockstar of a stonecutter, sliced the nodule for us.

CHAPTER 3: LIGHTBRINGER

45 fertility and love: Mircea Eliade, *The Forge and the Crucible: The Origins and Structures of Alchemy,* 2d ed., trans. Stephen Corrin (Chicago and London: The University of Chicago Press, 1956), 20, 21, 22, 102.

46 Sexuality was possessed by the ground: Eliade, *The Forge and the Crucible,* 38, 46, 48.

46 "the bowels of the Earth-Mother": Eliade, *The Forge and the Crucible,* 56.

46 "the heavenly smith": Eliade, *The Forge and the Crucible,* 30.

46 "treasures will be opened up": Eliade, *The Forge and the Crucible,* 47.

47 "At the slightest touch": Thao Ku, ca. 950, quoted in Joseph Needham, *Science and Civilization in China,* vol. 4, part 1 (1962; Cambridge: Cambridge University Press, 2004), 70.

47 served as Liebig's guide: Brock, *Justus von Liebig,* 99.

48 "records of warfare": W. Buckland, "On the Discovery of Coprolites, or Fossil Fæces, in the Lias at Lyme Regis, and in Other Formations," *Transactions of the Geological Society of London* (February 6, 1829): 235.

48 "an universal laboratory": Quoted in Gray, *Phosphates and Superphosphate,* 12.

48 In England, the cultivation of legumes increased: G. P. H. Chorley, "The Agricultural Revolution in Northern Europe, 1750–1880: Nitrogen, Legumes, and Crop Productivity," *The Economic History Review,* New Series 34, no. 1 (February 1981): 71, 89.

48 the highest levels of productivity: Jan Luiten van Zanden, "The First Green Revolution: The Growth of Production and Productivity in European Agriculture, 1870–1914," *The Economic History Review* 44, no. 2 (1991): 224.

48 "If the commercial manufacturing process": C. J. Dawson and J. Hilton, "Fertiliser availability in a resource-limited world: Production and recycling of nitrogen and phosphorus," *Food Policy* 36 (2011): S14–S22.

49 The yields of wheat in England: A. von Peter, "Fertiliser Requirements in Developing Countries," *Proceedings of the Fertiliser Society* 188 (1980): 13.

49 According to a 1935 analysis: M. K. Bennet, "British Wheat Yield per Acre for Seven Centuries," *Economic History* 3, no. 10 (February 1935): 28.

49 the application of artificial fertilizers in Britain: Van Zanden, "The First Green Revolution," 224.

49 Where fertilizers had been applied: Van Zanden, "The First Green Revolution," 219–20.

50 human waste in London: Lee Jackson, *Dirty Old London: The Victorian Fight Against Filth* (London: Yale University Press, 2014), 47.

50 John Martin proposed a centralized sewer system: Jackson, *Dirty Old London,* 66.

50 Edwin Chadwick: Jackson, *Dirty Old London,* 76.

50 the idea of fertilizer output: Jackson, *Dirty Old London,* 84.

51 Bazalgette proposed: Jackson, *Dirty Old London,* 162–63.

51 two thousand shiploads per year: Jackson, *Dirty Old London,* 103.

51 Joseph Priestley: Robert E. Schofield, *The Enlightened Joseph Priestley: A Study of His Life and Work From 1773 to 1804* (University Park: The Pennsylvania State University Press, 2004), 218, 229–30, 238–39.

52 In 1771, he began putting mint plants: Joseph Priestley, *Experiments and Observations on Different Kinds of Air,* 2d ed. (London: J. Johnson, 1775), 52.

52 He had already found that a mint plant: Priestley, *Experiments and Observations,* 50.

52 "repaired by the vegetable creation": Priestley, *Experiments and Observations,* 93.

53 Antoine Lavoisier: Madison Smartt Bell, *Lavoisier in the Year One* (New York and London: W. W. Norton, 2005).

54 "We may lay it down": See, for example: Trevor H. Levere, "Lavoisier," *Nature, Experiment, and the Sciences: Essays on Galileo and the History of Science* (Dordrecht: Kluwer Academic Publishers, 1990), 214.

55 The oldest known phosphate rock: For all episodes of phosphate formation, see:

Phosphate Deposits of the World (Cambridge University Press), vol. 1, *Proterozoic and Cambrian phosphorites,* ed. P. J. Cook and J. H. Shergold, 2005; vol. 2, *Phosphate Rock Resources,* ed. A. J. G. Nothold, R. P. Sheldon, and D. F. Davidson, 2005; vol. 3, *Neogene to Modern Phosphorites,* ed. William C. Burnette and Stanley R. Riggs, 2006. This section of the text is based on conversations with several geologists including John Compton of the University of Cape Town.

55 the Great Oxygenation Event: The geologist Nick Butterfield of Cambridge University talked me through the intricacies of this period in earth's history.

56 Phosphate fed the Cambrian explosion: P. J. Cook and J. H. Shergold, *Proterozoic-Cambrian Phosphorites: First International Field Workshop and Seminar on Proterozoic-Cambrian Phosphorites, held in Australia, August 1978* (Canberra: ANU Press, 1965–1991).

57 Some of that life was preserved in stone: Nothold, Sheldon, and Davidson, *Phosphate Deposits of the World,* vol. 2, 363.

57 In the northeast of Estonia: A.V. Ilyin and H. N. Heinsalu, "Early Ordovician shelly phosphorites of the Baltic Phosphate Basin," *Geological Society, London, Special Publications* 52 (1990): 253–59.

57 In central Tennessee: H. Thompson Straw, "Phosphate Lands of Tennessee," *Economic Geography* 17, no. 1 (January 1941): 93–104.

58 The deposit has been named Phosphoria: Nothold, Sheldon, and Davidson, *Phosphate Deposits of the World,* vol. 2, 53–58.

59 "an order succeeded by anarchy": Bell, *Lavoisier in the Year One,* part 4.

60 "In the remains of an extinct *animal* world": Quoted in Henslow, "Reminiscences of a Scientific Suffolk Clergyman," 22–23.

60 Following Suffolk in 1847 . . . quickly consumed following their discovery: Gray, *Phosphates and Superphosphate,* 22–35.

61 global phosphate rock production: Gray, *Phosphates and Superphosphate,* 36.

CHAPTER 4: STONES FROM PAST TIMES

75 "The remedy is obvious": Humphry Davy, *Elements of Agricultural Chemistry: A New Edition with Instructions for the Analysis of Soils and Copious Notes* (Glasgow: Richard Griffin and Company, 1844), 2.

75 "powdered charcoal": Davy, *Elements of Agricultural Chemistry,* 23.

75 "supplied by manure": Davy, *Elements of Agricultural Chemistry,* 2.

75 "a very long and difficult business": Gray, *Phosphates and Superphosphate,* 242.

76 "or build another barn": Lewis B. Nelson, *History of the U.S. Fertilizer Industry* (Muscle Shoals, Alabama: Tennessee Valley Authority, 1966), 26.

76 "came to look upon its author": Wyndham Miles, "Sir Humphrey Davie, the Prince of Agricultural Chemists," *Chymia* 7 (1961): 126.

76 "the immortal author": Quoted in Miles, "Sir Humphrey Davie," 133.

76 two American businessmen: *Superphosphate: Its History, Chemistry, and Manufacture,* United States Department of Agriculture and Tennessee Valley Authority, December 1964, 37–43.

76 Samuel Johnson embarked on a mission: *Superphosphate: Its History, Chemistry, and Manufacture,* 40.

76 "he hurried back": Thomas B. Osborne, "Biographical Memoir of Samuel William Johnson 1830–1909," *Biographical Memoirs,* vol. 7 (Washington, D.C.: National Academy of Sciences, July 1911), 208.

77 "he does not compile a receipt-book": Samuel Johnson, "No. III—Agricultural Science," *The Country Gentleman* (1853): 193.

77 "apostle of a cause": Osborne, "Biographical Memoir of Samuel William Johnson," 209.

77 Between 1850 and 1920, U.S. consumption: Nelson, *History of the U.S. Fertilizer Industry,* 100.

78 including shipments from Packard's: *Superphosphate: Its History, Chemistry, and Manufacture,* 51.

78 producers in America collected phosphate: Nelson, *History of the U.S. Fertilizer Industry,* 42. Five thousand tons of dried blood were produced as fertilizer in the United States in 1880, and that number rose substantially as time went on. In addition to those sources, fertilizer manufacturers turned to known standbys—manure, fish, cottonseed meal, and bones. The national market for bonemeal generally exceeded one hundred thousand tons annually around the turn of the twentieth century, while tankage was used as fertilizer in excess of two hundred thousand tons per year up until the early twentieth century.

78 They commenced mining in 1888: See, for example: *Superphosphate: Its History, Chemistry, and Manufacture,* 45.

79 "phosphate, phosphate, thou art a treasure": Arch Frederic Blakey, *The Florida Phosphate Industry* (Cambridge: The Wertham Committee, Harvard University, 1973), 28.

79 "It needs replenishing": "Message to Congress re Phosphorus Resources," speech file 1134, May 20, 1938.

80 Producers in Bone Valley: Blakey, *The Florida Phosphate Industry,* 90.

80 "closely guarded": George R. Mansfield, *Phosphate Resources of Florida* (Washington, D.C.: United States Government Printing Office, 1942), 4.

80 the Florida congressional hearing: *Phosphate resources of the United States: Hearings before the joint committee to investigate the adequacy and use of phosphate resources of the United States,* Congress of the United States, Seventy-Fifth Congress, third session, 924–25, 933.

83 "The oil industry has the capital": Blakey, *The Florida Phosphate Industry,* 101–3.

CHAPTER 5: RAPID CHANGE

87 The rezoned land surrounded: See, for example: "Resolution No. R110-113—Development Order Amendment and Related Operating Permit Amendments for Hillsborough County Consolidated Mines/Mosaic Fertilizer, LLC (DRI #263)," Hillsborough County, Florida, adopted August 10, 2010.

89 borrowing heavily to do so: John Ikerd, who began working as an agricultural economist in 1962, was one of the people urging farmers to change their approach. "We'd say, 'Okay, you need to do enterprise accounting—you need to account for each one of those enterprises separately,'" Ikerd told me in 2020. "You'd look at how profitable each one of those enterprises are on the farm. And then the enterprises that are more profitable—you'd look towards specializing. Doing more of those things and doing less of the others. This enterprise analysis was looking at the farm, these separate profit centers, like a big corporation would do." In the 1980s, when a series of recessions forced American farmers to the brink of economic disaster, Ikerd was still pushing for greater industrialization. "We were out working with these farmers, trying to help them save their farms," Ikerd said. "We'd have them

bring in their records. I could see that the farmers who were in the biggest trouble were the ones that had been following what we so-called experts had been telling them to do." The more he saw, the more he realized that his advice was causing problems—so he stopped giving that advice. "I changed over and said, 'Look, there's something fundamentally wrong with this model.' And then if you look back, all the problems that I saw with it can go right back to that industrial model. The problems that we saw with the farmers going broke was, in that case, farmers had to borrow a lot of money because this industrial system takes a lot of money for land and equipment and buildings. And so they borrow." The farmers were going into debt in order to buy the tools, including artificial fertilizers, required to participate in the new industrial system. Their profits were higher, but so was their risk. They were damaging their own ability to succeed. As the decade dragged on, Ikerd shifted his research entirely, from supporting the industrialization of agriculture to opposing it. His new papers contradicted his old papers. His new contacts disagreed with his previous ones. He lost invitations to publish in journals, to be a member of advisory councils and search committees, to serve in administrative positions. Because he no longer espoused the view that short-term profits were the main goal of agriculture, Ikerd told me, "people said, 'Well, he's not really an economist anymore.'"

90 funded by the wealth: Nick Cullather, *The Hungry World: America's Cold War Battle Against Poverty in Asia* (Cambridge: Harvard University Press, 2010), 54–56.

90 Mexican Agricultural Program: See: Tore C. Olsson, *Agrarian Crossings: Reformers and the Remaking of the US and Mexican Countryside* (Princeton: Princeton University Press, 2017), 73–97.

91 "from lower to higher levels of soil nutrients": Cullather, *The Hungry World,* 61.

91 a blueprint for the world: Cullather, *The Hungry World,* 68.

91 Land reform seemed to be the answer: Cullather, *The Hungry World,* 202–3.

91 Larger farms—corporate entities: Cullather, *The Hungry World,* 88–89.

91 Its leaders wanted to remake Asia: Cullather, *The Hungry World,* 163.

92 "literally thousands of traditional varieties": Gurdev S. Khush, "Green revolution: The way forward," *Nature Reviews: Genetics* 2 (October 2001): 815–22.

92 global yields surged: Cullather, *The Hungry World,* 233.

92 The harvest was striking: Von Peter, "Fertiliser Requirements in Developing Countries," 7–8, 18, 21.

93 Today, the majority of inputs: David Tilman, "The greening of the green revolution," *Nature* 396 (November 9, 1998): 211–12.

93 increased nineteen-fold: Nelson, *History of the U.S. Fertilizer Industry,* 243.

94 Armour Agricultural Chemical Company: Scott Hamilton Dewey, "'Is This What We Came to Florida For?': Florida Women and the Fight Against Air Pollution in the 1960s," *The Florida Historical Quarterly* 77, no. 4 (1999): 509.

95 "Brahma cattle began dying": Quoted in Blakey, *The Florida Phosphate Industry,* 108.

95 suffering from fluorosis: Blakey, *The Florida Phosphate Industry,* 108–9; E. R. Hendrickson, "Dispersion and Effects of Air Borne Fluorides in Central Florida," *Journal of the Air Pollution Control Association* 11, no. 5 (2012): 220–32.

95 *Life* magazine ran a two-page photo spread: Martin Schneider, "Air Pollution," *Life* (February 7, 1969): 46–47.

97 "By recovering by-product fluosilicic acid": Private letter shared with author.

97 One ton of phosphate rock: National Field Investigations Center—Denver, "Reconnaissance Study of Radiochemical Pollution from Phosphate Rock Mining

& Milling," Environmental Protection Agency, Office of Enforcement (December 1973), 2.

97 sixty times higher: E. G. Baker, H. D. Freeman, and J. N. Hartley, "Idaho Radionuclide Exposure Study—Literature Review," U.S. Environmental Protection Agency Office of Radiation Programs with the U.S. Department of Energy (October 1987): 1.1.

97 In 1979, the U.S. Department of the Interior reported: John W. Sweeney and Charles W. Hendry, *Minerals in the Economy of Florida* (Washington, D.C.: United States Department of the Interior, Bureau of Mines, 1979), 15.

97 Negev Desert: Robert Kelley and Vitaly Fedchenko, "Phosphate Fertilizers as a Proliferation-Relevant Source of Uranium," *Non-Proliferation Papers* 59, E.U. Non-Proliferation Consortium (May 2017): 5.

97 Egypt and Syria: Kelley and Fedchenko, "Phosphate Fertilizers as a Proliferation-Relevant Source of Uranium," 7.

97 when Saddam Hussein: Kelley and Fedchenko, "Phosphate Fertilizers as a Proliferation-Relevant Source of Uranium," 4, 6–7.

99 "The silence is deafening": Hagai Amit, "Potential Phosphate Mine Casts Toxic Shadow Over Bedouin Community," *Haaretz* (January 9, 2016).

99 seventeen former phosphate mines: John Miller, "EPA says Idaho mine violates law," *Casper Star Tribune* (June 27, 2009).

99 lead, radium, thorium: Freeman and Hartley, "Idaho Radionuclide Exposure Study—Literature Review," 3.14; National Field Investigations Center—Denver, "Reconnaissance Study of Radiochemical Pollution from Phosphate Rock Mining & Milling," 40.

99 As selenium concentration increases: Jordan Allen, "Health Impacts of Selenium Related to Phosphorous Mining," Carleton College (published online February 15, 2021).

99 Monsanto paid a $1.4 million fine: Miller, "EPA says Idaho mine violates law"; "Idaho Mining Company Agrees to Pay $1.4 Million Penalty to Settle Alleged Clean Water Act Violations," U.S. Department of Justice press release (April 2011).

100 Four months after Monsanto was fined: Vanessa Grieve, "Officials urge proper approach to mining phosphate in area," *Idaho State Journal* (October 6, 2011).

100 "I wouldn't know where I was": Gary O. Pittman, *Phosphate Fluorides Toxic Torts* (self-published, 2011), 43.

102 "with a bullhorn": Laura Baum, "Neighborhood Perceptions of Proximal Industries in Progress Village, FL," graduate thesis, University of South Florida, May 2016, 107.

103 residents of Progress Village resisted: Baum, "Neighborhood Perceptions of Proximal Industries in Progress Village, FL," 73.

104 On August 20, the commission approved the stack: Baum, "Neighborhood Perceptions of Proximal Industries in Progress Village, FL," 74.

104 Gary Lyman went on to look into the cancer risks: See, for example: Leonard A. Cole, *Element of Risk* (Washington, D.C.: AAAS Press, 1993), 48–50.

105 In 1988, he joined a study: H. Stockwell et al., "Lung Cancer in Florida: Risks Associated with Residence in the Central Florida Phosphate Mining Region," *American Journal of Epidemiology* 128, no. 1 (1998): 78–84.

105 In a 2003 internal report: "Memorandum: Summary of FL Dept of Health's Pre & Post Phosphate Mining Reports," United States Environmental Protection Agency

Region 4 (May 12, 2003); "Florida Phosphate Imitative 'Talking Points,'" United States Environmental Protection Agency Region 4 (July 1, 2003).

105 "an increased lung-cancer mortality rate": Robert Fleischer, "Lung Cancer and Phosphates," *Environment International* 8 (1982): 381–85.

106 increased lung cancer and leukemia mortality: James H. Yiin et al., "A study update of mortality in workers at a phosphate fertilizer production facility," *American Journal of Industrial Medicine* 59, no. 1 (January 2016): 12–22.

106 In 1975, the EPA announced preliminary results: *Mineral Commodity Profiles: Phosphate* (Washington, D.C.: Bureau of Mines, U.S. Department of the Interior, 1979), 15.

106 Polk County Health Department: Wayne King, "Florida Phosphate Pollution Stirs Alarm," *The New York Times* (June 24, 1976).

106 four potential pathways for radiation exposure: United States Environmental Protection Agency, Region 4, "Central Florida Phosphate Industry Environmental Impact Statement" (November 1978), 4.1.

106 "No activity of the phosphate industry": United States Environmental Protection Agency, Region 4, "Central Florida Phosphate Industry Environmental Impact Statement" (November 1978), 3.6–3.7.

108 It repeatedly violated: Michael Van Sickler, "Coronet says plant closing is just business," *Tampa Bay Times* (August 27, 2005).

109 as high as ninety-seven: *United States Environmental Protection Agency v. Coronet Industries* (docket no. CERCLA-04-2008-3755, Administrative Settlement Agreement and Order on Consent for Remedial Investigation/Feasibility Study), 6–8.

110 25 percent of all phosphate rock: The Mosaic Company, "Form 10-K (2020)," filed with the United States Securities Exchange Commission (February 22, 2021), 6–7. (In 2018, 41 percent of all phosphate rock produced in North America came from Four Corners [14.2 million tonnes in 2018 accounted for 56 percent of phosphate rock produced in North America; 6.2 of 14.2 million tons came from Four Corners].)

110 the Mosaic Company pointed to two local reports: A spokesperson for Mosaic declined—or ignored—requests for comment several times over a four-year period but referred to two studies, conducted by the Florida Department of Health and the Hillsborough County Environmental Protection Commission, in conjunction with Mosaic, that did not find a connection between mining activities and poor health outcomes. There is a long and well-documented history of corporate entities casting doubt on the science of environmental and behavioral cancer causes. It took governments until the 1960s, for example, to begin to regulate the tobacco industry, although scientists had known for decades that tobacco use causes cancer, and elevated incidences of lung cancer had been observed as early as the years that followed World War I—when smoking had been common among frontline troops. Even in the 1930s, the connection between smoking and lung cancer was clear enough that Hitler tried, and failed, to ban tobacco in Nazi Germany. In the United States, scientists repeatedly urged the government to accept the science, but tobacco companies were able to cast doubt on the connection. The waves of revelation that occurred in the latter part of the twentieth century—about tobacco, about food preservatives, about industrial pollutants and other environmental toxins—were really just resurgences of memories blocked for decades. Public concerns about asbestos, the popular building material, had begun in the 1920s,

and *Scientific American* wrote clearly about its carcinogenic properties in 1949. By that point, extensive research had shown the connections between asbestos and elevated cancer rates around the world, and conclusive research on the subject had been conducted in 1939. Yet the asbestos industry continued to push back against mounting evidence, making attempts to discredit research or alter the findings altogether. Beginning in the 1980s, the industry's efforts failed, and sixty-seven countries—including all members of the European Union—banned asbestos. "In the 1930s and 1940s, leading medical journals in Germany and a number of other industrializing countries frequently warned about the dangers of food preservatives, industrial toxins and coal-tar based artificial colorings while arguing for 'natural' products to be employed in drugs, cosmetics and foods," the epidemiologist Devra Davis wrote in *The Secret History of the War on Cancer*. "Those who confront the tobacco industry's successors and imitators need to remember that when organizations promoting 'sound science' have an economic interest in the prevention of science, no amount of proof may be enough."

CHAPTER 6: THE FLOOD

115 "Piney Point is what happens": Stephen King (@StephenKing), "Piney Point is what happens when you let the sharpies and hustlers have free rein. Money talks and the environment walks," Twitter (now X), April 3, 2021.

115 "are actually discharging radioactive waste": National Field Investigations Center—Denver, "Reconnaissance Study of Radiochemical Pollution from Phosphate Rock Mining & Milling," 25, 33.

117 the problems started early on: See, for example: Christopher O'Donnell, "Piney Point from 1966—present: On the edge of disaster," *Tampa Bay Times* (April 3, 2003); Bethany Barnes, Christopher O'Donnell, and Zachary T. Sampson, "Failure at Piney Point: Florida let environmental risk fester despite warnings," *Tampa Bay Times* (April 17, 2021); Craig Pittman, Julie Hauserman, and Candace Rondeaux, "Bending the rules at Piney Point, a $140-million mess," *Tampa Bay Times* (June 20, 2006).

120 HRK Holdings: Alex Harris and Ryan Callihan, "NY investor behind Piney Point ran hedge funds, a blueberry farm and string of Hooters," *The Miami Herald* (April 16, 2021); Chris Anderson, "Lawsuit questioned how hedge fund manager purchased Piney Point, site of wastewater leak," *Sarasota Herald-Tribune* (April 6, 2021); Burt A. Folkart, "Mellinger, Founder of Frederick's of Hollywood, Dies," *Los Angeles Times* (June 4, 1990).

120 "a man's point of view": Burt A. Folkart, "Mellinger, Founder of Frederick's of Hollywood, Dies."

121 Fursa's assets declined: *Claude Worthington Benedum Foundation vs. William Francis Harley, III, et al.*, case 2:12-cv-01386-NBF, United States District Court for the Western District of Pennsylvania (June 6, 2013), 25.

126 the extent of the damage: Zachary T. Sampson, "The large release earlier this year is estimated to have injected a year's worth of nitrogen into Lower Tampa Bay in about 10 days," *Tampa Bay Times* (August 19, 2021).

CHAPTER 7: PEAK AND VALLEY

133 "what is now North America": David Adams Leeming, *Creation Myths of the World* (Santa Barbara: ABC-CLIO, 2010), 217.

133 "the jeweled spear of Heaven": Leeming, *Creation Myths of the World,* 156.

133 "with which to build the world": Leeming, *Creation Myths of the World,* 25.

134 Toner's research describes: Jonathan D. Toner and David C. Catling, "A carbonate-rich lake solution to the phosphate problem of the origin of life," *Proceedings of the National Academy of Sciences* 117, no. 2 (December 30, 2019): 883–88.

138 "While not a single country recognizes": Lino Camprubí, "Resource Geopolitics: Cold War Technologies, Global Fertilizers, and the Fate of Western Sahara," *Technology and Culture* 56, no. 3 (July 2015): 676–703.

138 No one knows: The story of Moroccan hegemony contains one brief interruption. In 2023, a company called Norge Mining made a surprising announcement. According to its founder, Michael Wurmser—whom the company identifies as "an astute economist, geostrategist and entrepreneur"—Norge had found, in Norway, an amount of phosphate rock that would double the world's supply. The reserve, Wurmser said, totaled at least 70 billion tonnes (77 billion tons by U.S. measurement), and the implications were significant. With so much new mineable phosphate, Morocco's share of the world supply would decline from three-quarters to barely more than one-third. Europe would be a phosphate powerhouse again, harkening back to the industry's nineteenth-century roots. "A huge Norwegian phosphate rock find is a boon for Europe," *The Economist* reported. "'Phosphogeddon' Averted, Norway to the Rescue," another publication wrote. The business site Money Control was even more dramatic: "Norway's phosphate mine find to re-write the Western narrative," the website's headline read.

The phosphate deposit that Norge Mining described is located in a zone of hardened magma called the Bjerkreim-Sokndal Layered Intrusion, often referred to simply as the BKSK. More than a hundred years ago, the geological community took an interest in the BKSK; twenty-five years ago, geologists began to focus on one specific detail. Some layers in the intrusion contain around 10 percent concentrations of the phosphate mineral apatite. In the early 2000s, the Geological Survey of Norway examined the phosphate-rich layers and drilled into the rock to confirm the surface observations. The scientists who performed this work tried to interest miners in the opportunity. The Geological Survey published webpages, shared data, and talked to people at industry events, but with limited results. Then, in 2019, Norge Mining—founded a year earlier with a stated focus on sustainability—acquired the licenses to explore the BKSK. The company conducted magnetic surveys and drilled deep holes in order to determine the concentrations of economic minerals in the intrusion. In 2023, it shared with the world what it had found.

"Something appears to have gone wrong in a dialogue between a journalist and one of the investors during the summer of 2023. A newspaper article stated that the deposit contained seventy billion tonnes of phosphate. Even if this figure is considered to be the total phosphate-bearing rock volume, this is a highly speculative number that has not been confirmed by technical reports," Henrik Schiellerup, the director of resources and environment for the Geological Survey of Norway, told me in the fall of 2023. "Their public reports are about three and a half billion tonnes phosphate-bearing ore," he said. "Use their public numbers." Schiellerup was one of the geologists who confirmed the presence of apatite in 2004, and he spent years afterward encouraging prospectors to investigate it further. When Norge Mining acquired the license to explore parts of the BKSK, Schiellerup and his colleagues met with the company's geologists, sharing what they knew. Economically, he said, the potential of the deposit is still significant: 3.5 billion tonnes makes a big deposit. And from a geological standpoint, the formation is fascinating, giving insight into the complicated dynamics of magma underground.

The shape of the BKSK could be compared to that of a soup bowl or a trough. It was formed, 920 million years ago, when magma pushed upward into the earth's crust. This was a time of great upheaval in northern Europe: continents were thrust together, mountain ranges were constructed, and the BKSK was formed. As the intruding magma solidified, it left behind layers of valuable minerals—apatite, magnetite, and ilmenite—that came to extend across the whole formation. The layers are generally several hundred feet thick and pull down at the center of the bowl, as though weighted. At the time of its formation, the BKSK was located twelve miles underground, but, over the succeeding 900 million years, the land above it eroded away, exposing the top of the igneous intrusion to the air. Now, in southeast Norway, the top of the intrusion occupies an area of about ninety square miles and extends about three miles underground.

The challenge for the miners is first accessing the layers within the formation and then preparing the phosphate to be processed into fertilizer. It is an expensive proposition, to say the least. These are not Neil Davies's sedimentary structures, showcasing the history of the biological world. They are not the pebbles of the Peace or the vast layers underlying desert in Morocco. This is a sea of igneous rock, set beneath stunning Nordic vistas, cataloging dramas that took place miles underground nearly a billion years ago, all in lava. "It is—at the moment—hard to believe that deposit can be exploited with profit in the near future," Axel Müller, a geologist at the Oslo Natural History Museum, told me. "The deposit is large but has a low phosphorus content compared to other deposits. It is in the media simply because the new exploration license holder of the prospect wants to attract investors."

It would be exceedingly difficult—and energy intensive—to extract phosphate from the BKSK. Igneous rock is difficult to extract and hard to process. But it is also, in some ways, safer. Dangerous elements, such as cadmium and uranium, attach themselves to underwater crystals—like those that formed in Florida—but do not concentrate as heavily in magma. Moreover, the phosphate in the BKSK formed alongside minerals that are needed for electronics, which would be mined and sold as well. The energy-intensive nature of the mining would be allayed somewhat by the system of cheap and plentiful renewable energy that exists in Norway. There is also the matter of significance. "After COVID-19 and the outbreak of the war in Ukraine, with the disruption of fertilizer supplies, the prices of fertilizer went up, the price of food went up, and you started to see food insecurity. You really need fertilizers to secure food supplies for everyone," Jana Plananska, a consultant with Norge, told me. The company says its mine will open in 2028, and Plananska insisted this would be the case. She said that Norge intends to be a friendlier company, one more connected to the processes of nature—different from the other phosphate miners of the world. (She also confirmed that the company has found 3.6 billion tonnes of phosphate rock, or about 5 percent of the world's total, rather than the 70 billion tonnes that had earlier been announced.) "The sustainable production and use of fertilizers go hand in hand. I believe that industry plays an important role. Typically, if you're talking about production, it's the mining sector. If you're talking about use, it's farmers and the fertilizer industry. But the value chain really goes from the mine to the farmer and then to the plate. It's a chain—it should be a circle."

Henrik Schiellerup told me, "All exploration companies are optimists. Otherwise, they wouldn't do anything."

138 The global phosphate mining regime: Gray, *Phosphates and Superphosphate,* 72–80.

139 the cost of processing rises: P. K. Steen, "Phosphorus availability in the 21st century: Management of a non-renewable resource," *Phosphorus & Potassium* 217 (September–October 1998): 2–3.

139 "peak phosphorus": Dana Cordell, Jan-Olof Drangert, and Stuart White, "The story of phosphorus: Global food security and food for thought," *Global Environmental Change* 19, no. 2 (2009): 292–305.

139 "a thing of the past": Cordell, Drangert, White, "The story of phosphorus," 299.

139 Many phosphate deposits are located: David A. Vaccari, "Sustainability and the phosphorus cycle: Inputs, outputs, material flow, and engineering," *Environmental Engineer: Applied Research and Practice* 12 (Winter 2011): 5.

140 it reported to the Securities and Exchange Commission: The Mosaic Company, "Form 10-K (2020)": 7–8.

141 "Responding to his father's pain": *Kirkland et al. v. Mosaic Fertilizer, LLC et al.,* no. 8:14-cv-1715-24TGW, United States District Court for the Middle District of Florida, Tampa Division, 7.

142 "is unaccounted for": Shailesh K. Paten and Allen E. Schreiber, "Fate and Consequences to the Environment of Reagents Associated with Rock Phosphate Processing," Florida Institute of Phosphate Research (February 2001): 7–8.

145 "a total kill": Nelson, *History of the U.S. Fertilizer Industry,* 113–14.

145 Two billion gallons: Nelson, *History of the U.S. Fertilizer Industry,* 113–14, 248; George Lane Jr., "Peace River Knows No Peace," *The St. Petersburg Times* (December 16, 1971).

145 "gasping for oxygen": Fred McCormack, "Peace River: Victim of Pollution," *Sarasota Herald-Tribune* (December 8, 1971).

145 The river became unusable: The slime spills DeGaeta witnessed were just two examples of the effects of phosphate mining on the Peace River. The river had been the site of the region's first mining, in 1888, and it had started, incidentally, in DeSoto County. The mining had devastated the ecosystem. "A dozen years ago there was not a better fishing ground in the country for black bass than Peace river. Today there is scarcely a game fish in it owing to the pollution by the phosphate companies," the *Wauchula Advocate* reported in a 1916 article reprinted in the *DeSoto County News.* In 1920, a new swimming pool was opened in Arcadia for the convenience of "those who enjoy bathing and are now compelled to enjoy that sport in the dirty waters of the Peace River," the *DeSoto County News* reported ahead of the pool's June 4 opening. Two years earlier, when the pool was still under consideration, the newspaper had elaborated: "It was at first thought that the waters of Peace river could be utilized for this purpose, provided bath rooms and other needed facilities were provided for, but upon full investigation it is not believed that the water of this stream is sufficiently clean for bathing purposes, due to the fact that the phosphate mines of Polk county are contaminating this otherwise beautiful stream." In 1953, Ellis Lanquist, a graduate student in biology at the University of Florida, completed a master's thesis that studied the health of the Peace River. Lanquist set up thirteen collecting stations along the length of the river and its major

tributaries, and he investigated the water at each station on a quarterly basis for one year. He tested the chemical attributes of the water, including its acidity and dissolved oxygen, counted fish and other species, and recorded the speed and color of the water and the characteristics of the mud that lined the river's bottom. His paper is replete with reports of discharges of slime—the leaked contents of clay settling areas—and acid into the river. "As a result of failures of enclosing dams, slime occasionally enters the Peace River," Lanquist wrote in his thesis. He observed a slime discharge that coated the river in September. In January, when the river appeared to be like "yellow buttermilk," he was told that "a minor break at the Watson Mine had caused the release" of slime into the river. Following a dam break in May, turbidity—a measure of the lack of clarity of water—increased from less than 50 parts per million to more than 11,800, and the mollusks in the river died. "The larger specimens of snails and sphaerids were almost completely destroyed. Under the 2 to 3 in. layer of slime, the blackened remains of decomposing Chara were found, and the former creamy, yellow slime had been blackened below the surface. The unfavorable, probably near-septic, conditions caused by the decomposition of the vegetation produced the death of a greater number of Mollusca and other organisms [than] would have been killed otherwise." (See Ellis Lanquist, "Supplement II to Volume II: A Biological Survey of the Peace River Florida," *Peace and Alafia Rivers Stream Sanitation Studies 1950–1953* [Jacksonville, Florida: Florida State Board of Health], 12, 54, 58.)

148 "my mother still made mayonnaise": "Three Polk County Residents Develop a Unique Minerals Process," *Polk County Historical Quarterly* (September 2001): 7.

148 *Rhizosmilodon fiteae*: Steven C. Wallace and Richard C. Hulbert Jr., "A New Machairodont from the Palmetto Fauna (Early Pliocene) of Florida, with Comments on the Origin of the Smilodontini (Mammalia, Carnivora, Felidae)," *PLoS ONE* 8, no. 3 (March 2013): 107.

149 those with an eye for golf course design: Nelson, *History of the U.S. Fertilizer Industry,* 247.

149 "I think what separates us": Michael Croley, "Streamsong Founder Discusses the Golf Resort's Genesis and Speculates About a Future Course," *Robb Report* (March 14, 2018).

CHAPTER 8: THE END OF EVERYTHING

157 The End of Everything: This chapter is adapted from my 2019 undergraduate thesis at Princeton University, titled "The Sentencing of Nauru." It also contains material drawn from my 2019 article in *Focus on Geography,* titled "The Republic of Nauru and the Legitimacy of Existence," which itself was adapted from the thesis.

158 An amount approaching 100 million tons: *Certain Phosphate Lands in Nauru (Nauru vs. Australia),* vol. 1 of *Memorial of the Republic of Nauru,* International Court of Justice (April 1990), 84.

158 the richest in the world: Robert Trumbull, "World's Richest Little Isle," *The New York Times Magazine* (March 7, 1982).

158 Four-fifths of the island's surface: See, for example: Republic of Nauru, "Republic of Nauru Second National Communication" (submitted to the United Nations Framework Convention on Climate Change, December 2014), 22.

160 the country's projects turned desperate: See, for example: Philip Shenon, "A Pacific Island Nation Is Stripped of Everything," *The New York Times* (Decem-

ber 10, 1995); David S. Hilzenrath, "Tiny Island Shelters Huge Cash Flows," *The Washington Post* (October 28, 1999); Carl McDaniel and John Gowdy, *Paradise for Sale: A Parable of Nature* (Berkeley and Los Angeles: University of California Press, January 2000), 49; "Paradise well and truly lost," *The Economist* (December 20, 2001); Anne Davies and Ben Doherty, "Corruption, incompetence and a musical: Nauru's cursed history," *The Guardian* (September 3, 2018); "History of Forest Heights," *Forest Heights Homeowners Association* (accessed online: https://www.fhhoa.com/page/40965~845859/History-of-Forest-Heights).

160 Australian consumption of superphosphate increased: Gray, *Phosphates and Superphosphate,* 165.

161 107 kilograms per person: Von Peter, "Fertiliser Requirements in Developing Countries," 32.

161 by a factor of 4.5: Von Peter, "Fertiliser Requirements in Developing Countries," 29.

161 Under the Inca: Gregory T. Cushman, *Guano and the Opening of the Pacific World* (Cambridge, MA: Cambridge University Press, 2013), 8.

162 "British guano pirates": Cushman, *Guano and the Opening of the Pacific World,* 45.

162 The operation was supported: Cushman, *Guano and the Opening of the Pacific World,* 83.

162 the collection of sixty-six islands: Cushman, *Guano and the Opening of the Pacific World,* 80–81.

162 "vast crowds attended the opening": Cushman, *Guano and the Opening of the Pacific World,* 58.

163 Arundel set about: Cushman, *Guano and the Opening of the Pacific World,* 96.

163 On Fanning: A. G. Bell and A. E. Stephen, "Fanning Island," analysis of guano samples (November 1899). Accessed at the National Archives (U.K.), file reference DO 140/512.

163 On Washington: A. G. Bell and A. E. Stephen, "Washington Island—Falenikau Bed," analysis of guano samples (November–December 1899). Accessed at the National Archives (U.K.), file reference DO 140/512.

163 Pacific Phosphate Company: On Arundel's watch, company representatives sent back a series of missives describing far-flung islands. "Goram rises gradually to a moderate height. On all sides small streams descend," one report read. "About 100 yards from the beach at the principal village (Ondor) rises a wall of coral rock, perhaps 20 feet high. Above this wall is an undulating surface of coral which slopes downwards towards the interior." Aru Island "is low, but not as flat as it appears from the sea." On the Kei islands—reported then as Ke—"Light colored limestone rocks rise abruptly from the water several hundred feet. The rocks are everywhere broken into fantastic pinnacles, worn by the action of the weather into honeycombed surface and sharp points." E. Osborne, "Islands in the Malay Archipelago" (March 8, 1911). Accessed at the National Archives (U.K.), file reference DO 104/512.

163 Nor, Aliar, Dito, and St. Augustine: Letter sent from Pacific Phosphate Company representative in Nauru (November 5, 1911). Accessed at the National Archives (U.K.), file reference DO 140/515.

163 Walpole Island was mined: Description of Walpole Island (July 9, 1911). Accessed at the National Archives (U.K.), file reference DO 140/512.

163 Fanning and Washington: A. M. T. Duncan, "Re Fanning Island and Washington"

(December 11, 1907). Accessed at the National Archives (U.K.), file reference DO 140/512.

163 "Would you like to be a king": John T. Arundel, "Report No. 171 on visit to Fanning Island, per S.S. 'Ocean Queen,' Sept. 1st, 1909" (September 6, 1909). Accessed at the National Archives (U.K.), file reference DO 140/512.

163 The Pacific Phosphate Company acquired both islands: "Fanning Island," excerpt of an article from the *Evening Times* (March 25, 1912). Accessed at the National Archives (U.K.), file reference DO 140/512.

163 2.4 million tons of guano: "Prospectus," description of Washington and Fanning Islands. Accessed at the National Archives (U.K.), file reference DO 140/512.

163 Holbourne Island: C. P. Young, "Report on Queensland Phosphate Deposits, Holbourne Island" (August 3, 1944). Accessed at the National Archives (U.K.), file reference DO 140/577.

163 The mining spread across the Pacific: Gray, *Phosphates and Superphosphate,* 78–79.

164 "With Nauru shot to pieces": James P. Shortall, "The Phosphate Isle of Makatea," *The Mirror* (April 1946): 6–7, 70. Accessed at the National Archives (U.K.), file reference DO 140/512.

164 mining commenced: The first shipment of phosphate from Banaba left the island on September 5, 1900. The first Nauruan shipment—1,917 tons on July 6, 1907—was wrecked, though more soon followed. On both islands, digging had begun some time earlier. See Albert F. Ellis, *Ocean Island and Nauru* (Sydney: Angus and Robertson Limited, 1935), 99, 139.

164 On Banaba: William Lever of Lever Brothers, the precursor to Unilever, funded much of the early work on Banaba. As in Nauru, the island's land was effectively destroyed. In 1945, British authorities—who had overseen the mining—forced the indigenous population of Banaba to move to Rabi Island, Fiji, more than a thousand miles away.

164 "experimentation and exploitation": Epeli Hau'ofa, *We Are the Ocean: Selected Works* (Honolulu: University of Hawai'i Press, 2008), 46.

165 "Sea routes were mapped on chants": Hau'ofa, *We Are the Ocean,* 73.

165 "a once boundless world": Hau'ofa, *We Are the Ocean,* 34.

165 "the legitimacy of their existence": Hau'ofa, *We Are the Ocean,* 75.

166 The stories he wrote down: Timothy Detudamo, *Legends, Traditions & Tales of Nauru* (Suva: IPS Publications, University of the South Pacific, 2008; originally published in 1938). For the loss of Nauru's stories, also see: International Court of Justice, 90–93; Nancy Viviani, *Phosphate and Political Progress* (Honolulu: University of Hawaii Press, 1970), 9, 37–38; Wilhelm Fabricius, *Nauru 1888–1900* (Canberra: Division of Pacific and Asian History, Research School of Pacific Studies, Australian National University), 1992.

166 "the dream has been realized": Ellis, *Ocean Island and Nauru,* 209.

166 "The story of Ocean Island and Nauru": J. B. C., "Review: Ocean Island and Nauru: Their Story," *International Affairs* 16, no. 1 (January–February 1937): 151–52.

166 The white settlement of Australia: Robert Hughes, *The Fatal Shore: The Epic of Australia's Founding* (New York: Knopf, 1986).

166 Watkin Tench: Watkin Tench, *A Narrative of the Expedition to Botany Bay* (Melbourne: Text Publishing, 2009), 18, 49–50, 80; see also: Hughes, *The Fatal Shore,* 105.

167 "solipsism and despair": Hughes, *The Fatal Shore,* 467–68.

167 "a gallows with no rope": Hughes, *The Fatal Shore,* 484.

167 sentenced by John Stevens Henslow: Walters and Stow, *Darwin's Mentor,* 198.

167 Australia developed a unique identity: Donald Horne, *The Lucky Country* (Ringwood, Victoria: Penguin Books, 1964); Geoffrey Blainey, *The Tyranny of Distance* (Melbourne: Sun Books, 1966).

168 Immigration Restriction Act: "An Act to place certain restrictions on Immigration and to provide for the removal from the Commonwealth of prohibited Immigrants" (December 23, 1901). See Australian Federal Register of Legislation (legislation.gov.au).

169 International Health and Medical Services: Work carried out by IHMS has generated complaints. In 2015 Sarah Hanson-Young, a senator representing South Australia in the Parliament of Australia, reported IHMS to the Australian Federal Police after *Guardian Australia* published leaked documents showing that the company failed to properly vet its medical workers and misrepresented the standard of care it could provide on the islands. Behrouz Boochani, an Iranian journalist held in detention on Manus Island since 2013, called IHMS "obviously a criminal medical system" in 2016 following the death of an Iranian asylum seeker. A coroner said the death resulted from "errors" in his care on the island.

170 International SOS: Princeton University, which funded part of this research, is a client of International SOS.

173 "things happen": Daniel Hurst, " 'Things happen': Tony Abbott on sexual assault allegations in offshore detention," *The Guardian* (March 20, 2015).

174 "strategic business portfolio": "Ferrovial position on the Amnesty International report on the Nauru RPC," Ferrovial press release (April 4, 2017).

174 "even the most troubled patients": Carl Takei, "The East Mississippi Correctional Facility Is 'Hell on Earth,'" American Civil Liberties Union (March 5, 2018).

177 over two thousand of those incident reports: Nick Evershed et al., "The Nauru Files: The lives of asylum seekers in detention detailed in a unique database," *The Guardian* (August 10, 2016).

177 Revenues increased: Australian Government Department of Foreign Affairs and Trade, "Aid Program Performance Report" (2013–14, 2014–15, 2015–16, and 2016–17).

178 Eames and Law had been in Nauru: "Affidavit of Geoffrey Eames," District Court of Nauru, case numbers 20, 21, 22, and 24 of 2015, 18 of 2016, April 2017; Geoffrey Eames, "Geoffrey Eames resigns as Chief Justice of Nauru," Supreme Court of Nauru (March 13, 2014); Republic of Nauru, *Nauru Bulletin,* issue 74 (November 27, 2012), issue 76 (December 28, 2012), issue 13-2013/89 (August 22, 2013), issue 19-2013/95 (November 22, 2013).

179 "either completely ignorant": Tess Newton Cain, "Deafening silence on rule of law in Nauru," *The Interpreter* (March 21, 2014).

CHAPTER 9: OVERHAUL

185 "buckets on the end of a stick": Interview with Melanie Wright, North Wyke Farm (September 14, 2021).

187 "Is it not an outrage": Simón Bolívar, "Letter from Jamaica," 1815, trans. Lewis Bertrand, *Selected Writings of Bolivar* (New York: The Colonial Press, Inc., 1951), accessed online at Brown University Library.

187 three-quarters larger: Miguel Altieri and Clara I. Nicholls, "Scaling Up Agroecological Approaches of Food Sovereignty in Latin America," *Development* 41, no. 4 (December 2008): 474.

188 Aboriginal people inhabiting the landmass: Bruce Pascoe, *Dark Emu: Black Seeds: Agriculture or Accident?* (Broome, Australia: Magabala Books, 2014), 20–43.

188 "The unusual quality and friability of the soil": Pascoe, *Dark Emu,* 25.

190 Unrest grew worldwide: Tore C. Olsson, *Agrarian Crossings* (Princeton: Princeton University Press, 2017), 99.

191 "linked in memory with dictatorship": Cullather, *The Hungry World,* 242–44.

191 The land itself was degraded: Khush, "Green revolution: The way forward."

193 microbes create pore space in aggregates: Feng Wang et al., "Evolution of the transport properties of soil aggregates and their relationship with soil organic carbon following land use changes," *Soil and Tillage Research* 215 (January 2022).

195 the soil was healthy: A. E. Johnston, "Soil organic matter, effects on soils and crops," *Soil Use and Management* 2, no. 3 (September 1986): 97, 100. Also see: D. S. Jenkinson and D. S. Powlson, "The effects of biocidal treatments on metabolism in soil—V. A method for measuring soil biomass," *Soil Biology and Biochemistry* 8 (1976): 209–13.

196 Fungi, which can grow in acid: See: Jeff Lowenfels and Wayne Lewis, *Teeming with Microbes* (Portland: Timber Press, 2010). See also: Lynn Margulis and Dorian Sagan, *Microcosmos* (Berkeley and Los Angeles: University of California Press, 1986).

197 These fungi also process phosphate: Lowenfels and Lewis, *Teeming with Microbes,* 66.

197 When fewer are needed, fewer exist: Lowenfels and Lewis, *Teeming with Microbes,* 21–25.

197 in order for plants to grow: Dawson and Hilton, "Fertiliser availability in a resource-limited world," S14–S22.

197 many soils across Asia lack zinc: D. A. Forno, C. J. Asher, and S. Yoshida, "Zinc deficiency in rice," *Plant and Soil* 42 (1975): 551–63.

198 half of all farmland is deficient in zinc: Maha V. Singh, "Micronutrient Deficiencies in Crops and Soils in India," *Micronutrient Deficiencies in Global Crop Production,* ed. B. J. Alloway (Berlin: Springer, Dordrecht, 2008): 93.

198 A three-year greenhouse study: Shamie Zingore et al., "Multiple benefits of manure: The key to maintenance of soil fertility and restoration of depleted sandy soils on African smallholder farms," *Nutrient Cycling in Agroecosystems* 80 (2008): 267–82.

198 sulfur deficiency is a common problem: S. L. Tisdale et al., "Atlas of Sulfur Deficiencies," *Sulfur in Agriculture* (Madison, WI: American Society of Agronomy, Crop Science Society of America, and Soil Science Society of America, 1986), 295–322.

198 deaths from cardiovascular disease peaked: Andrea Rosanoff, "Changing crop magnesium concentrations: Impact on human health," *Plant Soil* 368 (2013): 139–53.

199 "dandelion greens": Jo Robinson, *Eating on the Wild Side: The Missing Link to Optimum Health* (New York: Little, Brown and Company, 2013).

201 a sharply lowered risk: See also: Daniel Lieberman, *The Story of the Human Body: Evolution, Health, and Disease* (New York: Pantheon, 2014).

201 more than fifty parishes: Archer, *By a Flash and a Scare,* 62.

201 "to palliate the evils": Walters and Stow, *Darwin's Mentor,* 208.

201 "I will never cart": Russell-Gebbett, *Henslow of Hitcham,* 30.

202 "his own potatoes": Jenyns, *Memoir of the Rev. John Stevens Henslow,* 54.

202 Lázaro Cárdenas, the president of Mexico: Raj Patel, "The Long Green Revolution," *The Journal of Peasant Studies* 40, no. 1 (2013): 7–8.

203 halve food insecurity: African countries today produce significant amounts of food that is exported to Europe and elsewhere while many African people go hungry. Although some people describe a lack of food production capabilities on the continent, the problem is actually economic. "Africans are hungry not because there is no food, but because they have no entitlement to food," Nick Cullather wrote (*Hungry World,* 269).

203 Marshaling its clout: Timothy A. Wise, "Failing Africa's Farmers: An Impact Assessment of the Alliance for a Green Revolution in Africa," Global Development and Environment Institute Working Paper No. 20-01 (July 2020), 10.

203 Abuja Declaration: "The Abuja Declaration on Fertilizers for an African Green Revolution," African Union (June 2011).

203 With AGRA setting the agenda: Wise, "Failing Africa's Farmers," 10.

204 an international group of agricultural organizations: Lena Bassermann et al., "False promises: The Alliance for a Green Revolution in Africa" (Bamako, Berlin, Cologne, Dar es Salaam, Johannesburg, Lusaka, Nairobi: Biodiversity and Biosafety Association of Kenya et al., July 2020): 25–26.

205 more than the amount that is mined: D. Cordell, J-O Drangert, and S. White, "The story of phosphorus: Global food security and food for thought," *Global Environmental Change* 19 (2009): 292–305; Vaccari, "Sustainability and the phosphorus cycle," 6.

206 three-fifths of all cereal crops: Vaclav Smil, "Phosphorus in the Environment: Natural Flows and Human Interferences," *Annual Review of Energy and the Environment* 25 (November 2000): 76.

206 used to make biofuels: "World fertilizer prices surge 200% in 2007, hurting the poor," *Mongabay* (February 20, 2008).

206 The Chinese government's fertilizer subsidy: Yuxuan Li et al., "An analysis of China's fertilizer policies: Impacts on the industry, food security, and the environment," *Journal of Environmental Quality* 42 (2013): 972–81.

206 In India: Reuters, "India faces record fertilizer subsidy in 2021–22," *The Hindu* (December 1, 2021).

206 Nor does it include Indonesia: Randy Fabi, "Subsidy sham: Fertilizers reach Indonesia plantations, not small farmers," Reuters (February 15, 2016); "Indonesia: Provided IDR 26.62 trillion fertilizer subsidy in the year 2020," *Global Trade Alert* (January 6, 2020).

207 an inverse relationship between the size of a farm: See, for example: Bladgely et al., "Organic agriculture and the global food supply," *Renewable Agriculture and Food Systems* 22, no. 2 (June 2007): 86–108.

207 about half of all food: A measure of calories produced yields a very different result than one of nutrients eaten. Large farms account for the majority of calories produced each year, but that number includes the world's production of nutrient-poor plants that go toward making animal feeds, biofuels, and other chemical concoctions. Small farms account for, perhaps—depending on the study methodology—a majority of the food that is actually consumed by people. Many different studies exist, and they make many different claims. Here, I split the difference.

207 The World Bank: Cullather, *The Hungry World,* 266.

207 countries in northwest Europe: B. Ulén et al., "Agriculture as a phosphorus source for eutrophication in the north-west European countries, Norway, Sweden, United Kingdom and Ireland: A review," *Soil Use and Management* 23, no. s1 (September 2007): 5–15.

208 The Netherlands has created a manure trading system: Qian Liu et al., "Global animal production and nitrogen and phosphorus flows," *Soil Research* 55 (2017): 458.

209 In Windhoek, the capital of Namibia: L. G. Moyo et al., "Evaluation of suitability of Windhoek's wastewater effluent for re-use in vegetable irrigation: A case study of Gammams effluent," *WIT Transactions on Ecology and the Environment* 199 (2005): 109–20; Christopher Maquet, "Wastewater reuse: A solution with a future," *Field Actions Science Reports* Special Issue 22 (2020): 64–69.

209 "West African agricultural repertoire": Dawit Solomon et al., "Indigenous African soil enrichment as a climate-smart sustainable agriculture alternative," *Frontiers in Ecology and the Environment* 14, no. 2 (March 2016): 71–76.

209 Chinampas are floating gardens: Altieri and Nicholls, "Scaling up Agroecological Approaches of Food Sovereignty in Latin America," 476.

210 "archaic and outdated": Kanok Rerkasem et al., "Land use transformation in the mountainous mainland Southeast Asia region and the role of indigenous knowledge and skills in forest management," *Forest Ecology and Management* 257 (2009): 2035–43.

210 "guano gardens": Cushman, *Guano and the Opening of the Pacific World*, 230–31.

CHAPTER 10: TINY TRACINGS ON A FUTURE WORLD

214 Dust from Asia: Andrew C. Kurtz, Louis A. Derry, and Oliver A. Chadwick, "Accretion of Asian dust to Hawaiian soils: Isotopic, elemental, and mineral mass balances," *Geochimica et Cosmochimica Acta* 65, no. 12 (2001): 1971–83.

214 Up to half the available phosphorus in the Sierra Nevada: S. M. Aciego et al., "Dust outpaces bedrock in nutrient supply to montane forest ecosystems," *Nature Communications* 8 (2017): 1–10.

215 State subsidies of phosphate rock–based fertilizers: H. Pathak et al., "Nitrogen, phosphorus, and potassium budgets in Indian agriculture," *Nutrient Cycling in Agroecosystems* 86 (2010): 287–99.

216 The South China Sea is eutrophic: R. Naylor et al., "Losing the Links Between Livestock and Land," *Science* 310 (2005): 1621–22.

216 forty-nine had eutrophied: W. Ma et al., "Phosphorus flows and use efficiencies in production and consumption of wheat, rice, and maize in China," *Chemosphere* 84 (2011): 814–21.

216 the global count of dead zones had doubled: Robert J. Diaz and Rutger Rosenberg, "Spreading Dead Zones and Consequences for Marine Ecosystems," *Science* 321 (2008): 927–29.

216 they experienced periods of oxygen deprivation: Katja M. Meyer and Lee R. Kump, "Oceanic Euxinia in Earth History: Causes and Consequences," *Annual Review of Earth and Planetary Sciences* 36 (2008): 267, 277.

217 the total annual flow of phosphate into the Baltic Sea: Smil, "Phosphorus in the Environment," 76.

217 Baltic Sea Action Plan: Accessed online. See: http://maps.helcom.fi/website/HELCOMexplorer/.

217 The southern third of the Baltic: Smil, "Phosphorus in the Environment," 76.

217 Researchers estimate that 30 percent: Diaz and Rosenberg, "Spreading Dead Zones and Consequences for Marine Ecosystems," 927.

218 "If the dead zones were eliminated": Diaz and Rosenberg, "Spreading Dead Zones and Consequences for Marine Ecosystems," 928.

218 phosphorus cycles through living bodies: David A. Vaccari, "Phosphorus: A Looming Crisis," *Scientific American* (June 2009): 54–59.

218 Vivianite, an iron-phosphate mineral: Matthias Rothe, Andreas Kleeberg, and Michael Hupfer, "The occurrence, soils and aquatic sediments identification and environmental relevance of vivianite in waterlogged soils and aquatic sediments," *Earth-Science Reviews* 158 (2016): 56.

219 phosphorus deposition generally corresponds to phosphorus flux: Karl B. Föllmi, "160 m.y. record of marine sedimentary phosphorus burial," *Geology* 23 (1995): 503.

219 SEDEX: Kathleen C. Ruttenberg, "Development of a sequential extraction method for different forms of phosphorus in marine sediments," *Limnology and Oceanography* 37, no. 7 (1992): 1460–82.

219 There is evidence of phosphate rock formation: Richard A. Jahnke et al., "The present day formation of apatite in Mexican continental margin sediments," *Geochimica et Cosmochimica Acta* 47, no. 2 (February 1983): 259–66.

219 Phosphate rock is forming in the Santa Barbara Basin: Christopher S. Martins, Robert A. Berner, and Jeffrey K. Rosenfeld, "Interstitial water chemistry of anoxic Long Island Sound sediments. 2. Nutrient regeneration and phosphate removal," *Limnology and Oceanography* 23, no. 4 (1978): 605–17.

220 Gabriel Filippelli: Filippelli, "The Global Phosphorus Cycle," 94.

221 the flow of phosphate from land into the oceans: Föllmi, "160 m.y. record of marine sedimentary phosphorus burial," 503–6.

223 "the rhythm of Time": Eliade, *The Forge and the Crucible,* 51.

223 "Life changes": Interview with Dan Canfield (June 15, 2020). "Phosphorus shouldn't be deemed a pollutant or a poison—it's a factor of life. You need it. Question is, what life will show up? Are you happy with it or not? Life changes as you get richer and richer."

224 A five-minute call: J. M. Langley, "The Phosphate Islands," *Fisons Journal* (April 1961): 27–31.

225 "an ultimate and unavoidable end": Richard Holmes, *The Age of Wonder: How the Romantic Generation Discovered the Beauty and Terror of Science* (New York: Pantheon, 2008), chapter 4.

225 Astronomers were looking for phosphorus in the heavens: A team led by Jane Greaves at the University of Cardiff.

INDEX

Page numbers in *italics* refer to illustrations.

A NOTE ABOUT THE TYPE

This book was set in a version of the well-known Monotype face Bembo. This letter was cut for the celebrated Venetian printer Aldus Manutius by Francesco Griffo, and first used in Pietro Cardinal Bembo's *De Aetna* of 1495.

The companion italic is an adaptation of the chancery script type designed by the calligrapher and printer Ludovico degli Arrighi.

Composed by North Market Street Graphics
Lancaster, Pennsylvania

Printed and bound by Berryville Graphics
Berryville, Virginia

Designed by Marisa Nakasone